AF389003

Lifetime Prediction and Simulation Models of Different Energy Storage Devices

Lifetime Prediction and Simulation Models of Different Energy Storage Devices

Special Issue Editor

Julia Kowal

MDPI • Basel • Beijing • Wuhan • Barcelona • Belgrade

Special Issue Editor
Julia Kowal
Technical University of Berlin,
Electrical Energy
Storage Technology
Germany

Editorial Office
MDPI
St. Alban-Anlage 66
4052 Basel, Switzerland

This is a reprint of articles from the Special Issue published online in the open access journal *Sustainability* (ISSN 2071-1050) from 2019 to 2020 (available at: http://www.mdpi.com).

For citation purposes, cite each article independently as indicated on the article page online and as indicated below:

LastName, A.A.; LastName, B.B.; LastName, C.C. Article Title. *Journal Name* **Year**, *Article Number*, Page Range.

ISBN 978-3-03936-561-6 (Hbk)
ISBN 978-3-03936-562-3 (PDF)

Contents

About the Special Issue Editor

Julia Kowal studied Electrical Engineering and Information Technology at RWTH Aachen University. She also conducted her Ph.D. in Electrical Engineering here within the Institute for Power Electronics and Electrical Drives (ISEA), Electrochemical Energy Conversion, and Storage Systems on the topic of "Spatially Resolved Impedance of Nonlinear Inhomogeneous Devices—Using the Example of Lead-Acid Batteries". After finishing her Ph.D., she worked as Chief Engineer and Head of the department "Modelling, Analytics and Lifetime Prediction", Electrochemical Energy Conversion and Storage Systems, ISEA, RWTH Aachen University. She has served as Professor of Electrical Energy Storage Technology at TU Berlin since her appointment in 2014. Her research focuses on the characterization and modelling of different battery technologies, such as lithium-ion batteries, lead–acid batteries, and metal–air batteries with the aim of lifetime prediction and optimized operation as well as state determination.

Preface to "Lifetime Prediction and Simulation Models of Different Energy Storage Devices"

Energy storage is one of the most important enablers for the transformation to a sustainable energy supply with greater mobility. For vehicles, but also for many stationary applications, the batteries used for energy storage are very flexible but also have a rather limited lifetime compared to other storage principles. Many different battery technologies exist which, in part, show a similar aging behavior, but each of which mostly has their own characteristic aging processes. Other storage principles such as supercaps, fuel cell/electrolyzer/storage systems, or thermal storage also show declining performance with time. To develop reliable systems and facilitate cost calculations, it is therefore necessary to understand the aging processes as well as its influencing factors. This Special Issue is a collection of articles that collectively address the following questions:

- What are the factors influencing the aging of different energy storage technologies?

- How can we extend their lifetime? How can we optimize the operation of energy storage for the optimum lifetime, while fulfilling the purpose of storage?

- How can the aging of an energy storage be detected and predicted? When do we have to exchange the storage device?

  The articles deal with different topics in the following fields:

- Aging simulation models of different energy storage devices

- State of health detection of different energy storage devices

- Lifetime tests and analysis of influencing factors of different energy storage devices

- Operating strategies with the aim of an optimized lifetime for different energy storage devices

Julia Kowal
Special Issue Editor

Article

Study of a Li-Ion Cell Kinetics in Five Regions to Predict Li Plating Using a Pseudo Two-Dimensional Model

Sanaz Momeni Boroujeni * and Kai Peter Birke

Electrical Energy Storage Systems, Institute for Photovoltaics, University of Stuttgart, Pfaffenwaldring 47, 70569 Stuttgart, Germany; Peter.Birke@ipv.uni-stuttgart.de
* Correspondence: sanaz.momeni@ipv.uni-stuttgart.de

Received: 17 October 2019; Accepted: 7 November 2019 ; Published: 14 November 2019

Abstract: Detecting or predicting lithium plating in Li-ion cells and subsequently suppressing or preventing it have been the aim of many researches as it directly contributes to the aging, safety, and life-time of the cell. Although abundant influencing parameters on lithium deposition are already known, more information is still needed in order to predict this phenomenon and prevent it in time. It is observed that balancing in a Li-ion cell can play an important role in controlling lithium plating. In this work, five regions are defined with the intention of covering all the zones participating in the charge transfer from one electrode to the other during cell cycling. We employ a pseudo two-dimensional (P2D) cell model including two irreversible side reactions of solid electrolyte interface (SEI) formation and lithium plating (Li-P) as the anode aging mechanisms. With the help of simulated data and the Nernst–Einstein relation, ionic conductivities of the regions are calculated separately. Calculation results show that by aging the cell, more deviation between ionic conductivities of cathode and anode takes place which leads to the start of Li plating.

Keywords: li-ion cells; lithium plating; kinetic balancing; ionic conductivity; modeling

---

1. Introduction

Lithium ion (Li-ion) batteries were first developed in the 1970s [1–3]. After two decades of intensive materials development, Li-ion cells were commercialized by Sony in 1991 [4,5]. Constructed as the best compromise due to many excessive failures of rechargeable Li-metal cells beforehand, Li-ion cells have undergone a tremendous evolution in the last few decades and have been widely utilized for energy storage in different portable, computing, and telecommunicating devices as well as electrified transport vehicles. Clean electric conveyances are a possible way to reduce the environmental impact of private transport and abate about 10–20% of the emissions [6,7]. However, increasing the energy density of Li-ion batteries to accomplish the actual demand of electrified vehicles is of importance. As a solution, electrodes have to become thicker and denser. If the negative carbon-based electrode is manufactured with a higher thicknesses, more density, and less electrolyte uptake, then the occurrence of Li-plating becomes an unavoidable issue. Moreover, capacity retention, lifetime, fast and low temperature charging, and safety performance of the cells still require improvements. These challenging demands are all directly or indirectly influenced by lithium deposition [8–10]. The appearance of metallic lithium on the surface of carbon particles is a complex function of temperature, aging, and cycle loads.

Lithium deposition or lithium plating, which generally means the formation of metallic lithium on the negative electrode is an all-time undesirable phenomenon, contributing to cell performance degradation, reducing the cell durability and cyclability. Additionally, it significantly raises safety issues [11]. Avoiding or suppressing the lithium deposition reaction is essential to the reliability and also improvement of Li-ion cells. This paper introduces a way to investigate plating occurrence. Therefore, it is important to describe the phenomenon of Li-Plating in more details.

While charging, lithium ions are extracted from the cathode, diffuse through the electrolyte, and intercalate into the graphite structure of the negative electrode. The Li^+ charge transfer process (CTP) takes place on the surface of electrode particles, meaning the de-solvation of solvated lithium ion from the electrolyte, transferring through the solid electrolyte interface (SEI) layer and entering into the electrode particle, becoming an intercalated lithium. The intercalated lithium diffuses further inside the insertion electrode, i.e., the graphene planes of graphite, preventing the particle surface to reach saturation. This is called the lithium solid diffusion process (SDP). Depending on the operation conditions e.g., at a high C-rate, low temperature and high state of charge (SOC), the CTP and SDP will be the limiting factor, respectively. It means the speed of the lithium ion flow in the electrolyte is exceeding the charge transfer process or solid diffusion. Consequently, the lithium deposition reaction can occur instead or in parallel with intercalation. Thermodynamically, lithium plating is not favorable in comparison to intercalation since its reaction enthalpy is more positive [12,13]. Nevertheless, during a charge process, due to the deviation from equilibrium, an induced overpotential is formed [14] which may cause the anode potential to drop below 0 V vs. Li^+/Li and consequently the lithium deposits on the surface of the graphite particle [15]. It should be noted that for intercalation, the potential range is from 200 to 65 mV vs. Li^+/Li [13]. Lithium solid diffusion overpotential and charge transfer overpotential, which enable the lithium deposition reaction, are the kinetic causes of this phenomenon. Operating conditions like low temperature (below 25 °C), fast charging, and high SOCs increase the cell overpotentials [16,17].

According to the literature [18], a Li^+ ion which is transferred from electrolyte to the anode experiences discrete energy barriers at different regions. Diffusing in the electrolyte (considered a liquid) has a relatively low energy barrier while charge transfer through the SEI usually has the biggest activation energy (E_a) and diffusion of Li into graphite has a moderate energy barrier in the range of 0.22–0.4 eV, increasing by x in Li_xC_6 [19,20]. According to Arrhenius, a low temperature slows down the reactions [14] which means less Li^+ ions can overcome the charge transfer energy barrier to intercalate. At the same time, the solid diffusion of Li happens more sluggish at high SOCs. Therefore, when the charging current is high enough to induce a big ion flux toward the anode, so that the Li solid diffusion rate cannot compete, lithium deposition happens with a higher probability [15,17,21].

There is some literature investigating factors influencing CTP kinetics. Some believe that the Li^+ de-solvation step is always slower than Li^+ transferring through the SEI [22,23]. Some other investigate different electrode as well as electrolyte materials and conclude that the kinetics of Li^+ charge transfer process is controlled by the chemistry of the electrode components and their interfacial layer. By having a SEI layer with a low energy barrier for conducting Li^+, the de-solvation step is limiting and vice versa if the SEI layer is not conductive enough [24].

Numerical simulations provide quantitative information to further investigate the phenomenon. Firstly Arora et al. [25] introduced a numerical way for describing and predicting the lithium deposition at charge and overcharge. Years later Tang et al. [26] extended the previous work to 2D. Legrand et al. [13] investigated Li-P (lithium plating) through CTP limitations by an electrochemical model, yet they have not examined the SPD process. Jiang et al. [27] proposed characteristic times to explain charge and species transport limitations in Li-ion batteries, but they did not cover the aging influences. Understanding the lithium plating phenomenon has been the focus of many studies, however there is still a lack of information on the mechanisms of transport-related performance limitations during charge/discharge operations over the life time of Li-ion cells. In the present model-based study we investigate the transport mechanisms behavior of Li-ion cell over 400 cycles. Aging mechanisms are included by a growing surface layer consisting of SEI and plated lithium. We introduce and explain five regions in the cell (A-E) which are contributing to the Li^+ charge transfer and Li solid diffusion processes. Ionic conductivities of these regions are calculated afterwards. The cell aging behavior and appearance of Li plating from the 116th cycle of the simulated cell are explained and discussed with the help of ionic conductivity variation in the mentioned regions.

The paper is organized as follows. Section 2 explains the transport regions and implementation of our P2D (pseudo two-dimensional) model. In Section 3 we show the validation of our calculations, discuss the model results, and explain them with the help of ionic conductivity calculations. Section 4 summarizes the results.

2. Theory

2.1. Transport Regions

In addition to electrical conduction, ionic conductivity in the electrodes and electrolyte is necessary for the completion of electrochemical reactions. Charged species, including Li-ions can pass through a media under two driving forces: An externally applied electric field and/or a concentration gradient which is described in the Nernst–Planck relation:

$$j_{ion} = -c_{ion}v_{ion} = \frac{u_{ion}}{z_{ion}q}kT(\nabla c_{ion} + \frac{F}{kT}c_{ion}\nabla\phi) = D_{ion}(\nabla c_{ion} + \frac{F}{kT}c_{ion}\nabla\phi) \tag{1}$$

where j_{ion} is the ionic current density, v_{ion} is the drift velocity, u_{ion} is the electrical mobility of ions, z_{ion} is the valence number, and q is the charge. K is the Boltzamn constant, D_{ion} is the diffusion coefficient, and $\nabla\phi$ is the gradient of potential. Ohm's law is the relation between current density (i), conductivity (σ), and the electric field (ζ). By substituting the ionic current density (j) for i into Ohm's relation and including chemical potential in driving forces, as it is needed for the ions, then rewriting the derived formula for σ_{ion} and finally comparing it with Nernst–Plank relation we come to an equation called the Nernst–Einstein:

$$\sigma_{ion} = \frac{c_{ion}D_{ion}z_{ion}^2F^2}{RT} . \tag{2}$$

During charge and discharge, Li$^+$ transfers from one side to the other. To study the transport mechanisms in the cell we consider five regions (see Figure 1), which can be defined as follows:

- Regions A and E representing the inner part of solid active material particles in the anode and cathode respectively;
- Regions B and D that are the solid particles/electrolyte interfaces at the anode and cathode side respectively;
- Region C which indicates the electrolyte which can be in the anode, the separator, or the cathode.

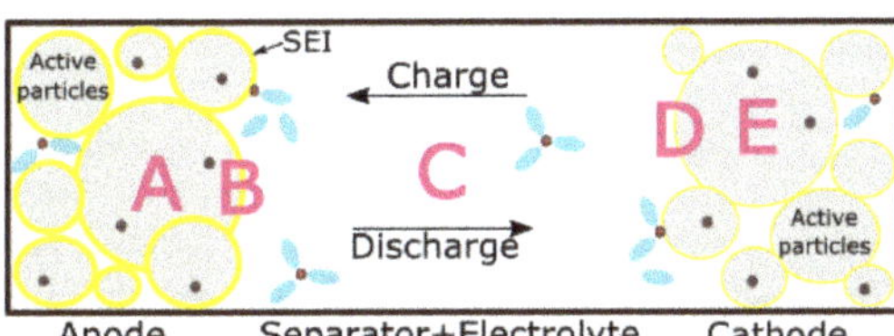

Figure 1. Schematic view of the five transport regions considered in a full lithium ion cell.

During the charge period, intercalated Li diffuses from the cathode particle bulk to the interface (E). Particles of the solid active materials at the cathode and anode are assumed to be spherical. At the surface, it donates one electron and crosses the formed SEI layer on particles at the interface (D) to enter the electrolyte. This electrochemical reaction can be explained by the Butler–Volmer equation. Li$^+$ diffuses further in the electrolyte towards the anode because of the concentration gradient and electric field (C). By reaching the anode particles, the Li$^+$ ion transfers through the surface layer formed on the particles and receives one electron (B) and enters into the particle. Intercalated lithium diffuses away from the surface due to the concentration gradient (A). Once the cell is fully charged and the discharge starts, the whole transport process takes place in a reverse direction from the anode side to the cathode.

Calculating the ionic conductivity for each of the five regions can give us a simple way to compare the transport performances at different conditions. In this work we calculate ionic conductivity of transport for (1) solid active materials A and E, (2) electrolyte C, and (3) interfaces B and D based on the Nernst–Einstein relation.

2.2. Model Description

The required data for calculation comes from an implemented model which is based on the pseudo two-dimensional (P2D) approach [28]. We used COMSOL Multiphysics® 5.4 software for simulation. A detailed explanation of the governing partial differential equations can be found in the literature [28–31]. The model considers charge and species transport along the electrodes thicknesses direction (x) and in the solid particles (r) of active materials. Equations governing the x and r directions are coupled via the electrochemical reactions on the surface of active material particles described by the Butler–Volmer relation. Lithium plating and SEI formation are considered aging mechanisms, so that they are the anode side reactions competing the intercalation reaction during the charge process. This means:

$$j_{tot} = j_{int} + j_{SEI} + j_{LiP} \tag{3}$$

where j_{int} is the intercalation current density, j_{SEI} is the current density of the SEI formation, and j_{LiP} is the current density of lithium plating. From the Butler–Volmer relation for the intercalation current density we have the definition of:

$$j_{int} = j_{0,int}\left(exp\left(\frac{\alpha_a F \eta_{int}}{RT}\right) - exp\left(\frac{-\alpha_c F \eta_{int}}{RT}\right)\right) \tag{4}$$

where $j_{0,int}$ is the intercalation exchange current density, α_a and α_c are anodic and cathodic transfer coefficients respectively. Exchange current density for intercalation can be calculated as follows:

$$j_{0,int} = F k_c^{\alpha_a} k_a^{\alpha_c} (c_{s,max-c_s})^{\alpha_a} (c_s)^{\alpha_c} \left(\frac{c_l}{c_{l,ref}}\right)^{\alpha_a} \tag{5}$$

k_a and k_c are the rate constants of the anodic and cathodic reactions respectively. The maximum possible concentration is $c_{s,max}$, and c_s is the local concentration of solid particles. c_l represents the electrolyte concentration. SEI formation, which we assumed in this model is the reaction of ethylene carbonate (EC) from electrolyte with lithium ions and electrons. A detailed description about the simulation of the SEI layer can be found in previous works [32,33]. The surface overpotential for each of the reactions (intercalation, SEI, Li-plating) is:

$$\eta_{(int,SEI,LiP)} = \phi_s - \phi_l - E_{eq(int,SEI,LiP)} - \Delta\phi_{film} \tag{6}$$

where ϕ_s and ϕ_l are the potential of solid and electrolyte phase respectively. $\Delta\phi_{film}$ is the potential drop over the film which is forming because of the SEI and Li-plating side reactions and E_{eq} is the equilibrium potential of the corresponding reaction.

Both side reactions are assumed to be irreversible. The additional oxidation of plated lithium and consequently the formation of the secondary SEI layer on the plated lithium is neglected. Additionally, it is assumed that no partial dissolution of deposited Li during the discharge occurs. The current density of each of the side reactions is calculated, considering only the cathodic part of the Butler–Volmer relation [34]:

$$j_{(SEI,LiP)} = -j_{0(SEI,LiP)}exp\left(\frac{-\alpha^c_{(SEI,LiP)} F \eta_{(SEI,LiP)}}{RT}\right) \tag{7}$$

Cell parameters originate from an experimental cell. Key cell parameters and simulation conditions are listed in Table 1. Diffusion coefficient of electrodes are adjusted based on Cabanero's work [35] to include a degree of lithiation dependency. The cycling is simulated using a constant current/constant voltage (CC/CV) charging strategy. Discharge is simulated using CC only. Since no thermal modeling is included, a 0.5 C charge/discharge rate is applied for the whole simulation so that the temperature variation over charge and discharge can be neglected [36].

Table 1. Cell parameters and simulation conditions used in the model.

Parameters	Anode	Separator	Cathode
Thickness L (μ_m)	116	16	88.7
Particle radius r_0 (μ_m)	8.8	-	6.5
Porosity ϵ (%)	0.26	0.5	0.24
Bruggeman exponent γ (-)	1.8	1.5	1.56
Initial electrolyte concentration c_e (mol/m^3)		1200	
Diffusion coefficient in solid D_s (m^2/s)	Figure 2A [35]	-	Figure 2B [35]
Diffusion coefficient in liquid D_l (m^2/s)		3.7×10^{-9} *	
Ionic conductivity in liquid σ_l (S/m)		8.735×10^{-1} *	
Maximum Li$^+$ concentration in solid (mol/m^3)	27,880	-	48,580
Anodic/Cathodic transfer coefficient α_a, α_c (-)	0.5		0.5
Transference number t_+ (-)		0.577	

Simulation conditions	Values
Temperature T (K)	298.15
Lower and upper voltage boundary $U_{min} - U_{max}$ (V)	2.7–4.2
Charge and discharge rate $C - rate$	0.5
Equilibrium potential SEI formation $E_{eq,SEI}$ (V)	0.4 [37]
Equilibrium potential lithium plating $E_{eq,LiP}$ (V)	0
Cathodic charge transfer coefficients for side reactions $\alpha^c{}_{SEI.LiP}$ (/)	0.5 [34]

* for c_l = 1200 mol/m^3.

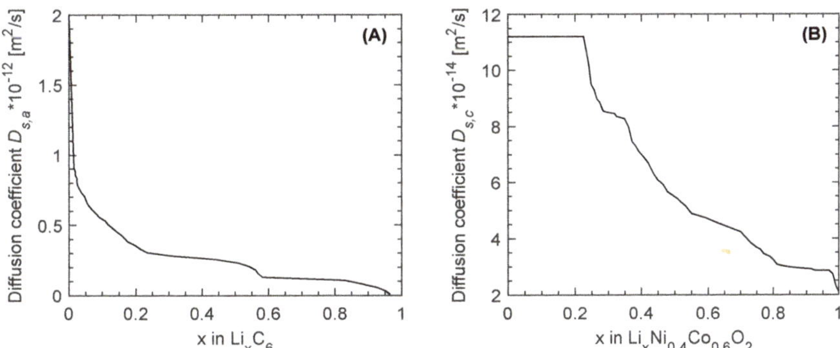

Figure 2. Diffusion coefficient (**A**) of the anode and (**B**) of the cathode as a function of the lithiation degree.

Based on the Nernst–Einstein relation, we defined ionic conductivity for each of the regions as follows:

$$\sigma_{l(a,c,s)} = \frac{c_l D^{eff}_{l(a,c,s)} F^2}{RT} \tag{8}$$

σ_l is the ionic conductivity of electrolyte (region C). D^{eff}_l is the effective diffusion coefficient of electrolyte which is defined depending on its medium i.e., anode, cathode, or separator. It is calculated with the Bruggeman correlation. So that:

$$D^{eff}_{l(a,c,s)} = D_l \epsilon^{\gamma}_{l(a,c,s)} \tag{9}$$

consisting of the electrolyte diffusion coefficient D_l, the electrolyte volume fraction $\epsilon_{l(a,c,s)}$, and the Bruggeman exponent γ.

For the particles of solid active material (region A and E) we have:

$$\sigma_{s(a,c)} = \frac{(c_{s,max(a,c)} - c_{s,ave(a,c)})D_{s(a,c)}F^2}{RT} \tag{10}$$

in which D_s is the diffusion coefficient of the solid particles.

To calculate the ionic conductivity for the last two regions (B and D) we assume the interface reactions to consume Li ion from electrolyte at electrode/electrolyte interface. Therefore ionic conductivity is:

$$\sigma_{i(a,c)} = \frac{c_{s(a,c)}\left(\frac{c_l}{c_{l-ref}}\right)^{\alpha_a}F^2}{RT} \tag{11}$$

including surface and electrolyte concentrations (c_s and c_l) coming from the Butler–Volmer relation definitions and effective electrolyte diffusion coefficients.

3. Results

3.1. Validation

To validate the ionic conductivity results coming from our calculations we compared them with the characteristic time values for transport which are introduced by Jiang and Peng [27]. They have defined three parameters, t_s, t_i, and t_l as:

$$t_{s(a,c)} = \frac{(r_{0(a,c)}/3)^2}{D_{s(a,c)}} \tag{12}$$

$$t_{i(a,c)} = \frac{Fec_l}{(1 - t_+^0)|j^{Li}|} \tag{13}$$

$$t_l = \frac{L_a^2}{D_{l,a}^{eff}} + \frac{L_s^2}{D_{l,s}^{eff}} + \frac{L_c^2}{D_{l,c}^{eff}}. \tag{14}$$

t_s is describing a characteristic time of the Li diffusion process into solid particles in negative and positive electrodes. t_i stands for the transport time relating to the local depletion rate of Li ions in electrolyte at the electrode/electrolyte interface, and t_l is the characteristic time for Li ion transport through the electrolyte. Considering these definitions we can relate $\sigma_{l(a,c,s)}$, $\sigma_{s(a,c)}$, and $\sigma_{i(a,c)}$ to $t_{l(a,c,s)}$, $t_{s(a,c)}$ and $t_{i(a,c)}$ respectively.

Using the cell parameters reported in the Jiang's article for simulation, we gain the following results for transport times and ionic conductivity calculations in idle state prior to discharge as listed in Table 2. There are slight differences between anode transfer time coming from our calculations and the one reported in Jiang's work. The reason might be (1) due to the differences in parameters assumptions as not all the values are mentioned in the article and (2) in contrast to Jiang's model we included Li-plating and SEI formation (anode aging mechanisms).

Table 2. Values of the characteristic times and corresponding ionic conductivities when the battery is in the pause state prior to discharge. Lit. values are extracted from Jiang's article [27].

	t_l [s]		σ_l [S/cm]	t_s [s]		σ_s [S/cm]	t_i [s]		σ_i [S/cm]
	Lit.	Cal.	Cal.	Lit.	Cal.	Cal.	Lit.	Cal.	Cal.
Anode		174	1.12×10^{-2}	3.2×10^3	2.2×10^3	$(3.19\text{-}2.98) \times 10^{-7}$	147.7	130	2.88×10^{-4}
Cathode	180	103.2	1.21×10^{-2}	71.1	71.1	1.28×10^{-4}	101.3	92.3	1.16×10^{-3}
Separator		8.8	2.15×10^{-2}	-	-	-	-	-	-

Comparing the ionic conductivities with their corresponding transport times, we realized that t and σ values of similar regions in positive and negative electrodes are following the same trend. This compatibility of results suggests that it is valid to compare $\sigma_{s,a}$ to $\sigma_{s,c}$ and additionally $\sigma_{i,a}$ to $\sigma_{i,c}$. As shown in Table 2, there is only one value reported for the transport time of electrolyte. Therefore it is not possible to check the trend of our discrete values of electrolyte ionic conductivity in different mediums with the transport time.

3.2. Simulation Results

The discharge capacity behavior of the simulated Li-ion cell with the mentioned parameters in Table 1 over the cycle number is shown in Figure 3. The relative discharge capacity is defined as the relation of current Q_{dis} to the first cycle discharge capacity. During the initial cycles, the discharge capacity decreased faster than the following cycles, which was when the SEI layer initially formed. The almost linear decrease continued until cycle number 230, where the Q_{dis} reached 78% of the initial capacity. Then the phase of nonlinear decrease in discharge capacity starts so that in total cycle numbers of 400, the cell lost more than 60% of its initial capacity.

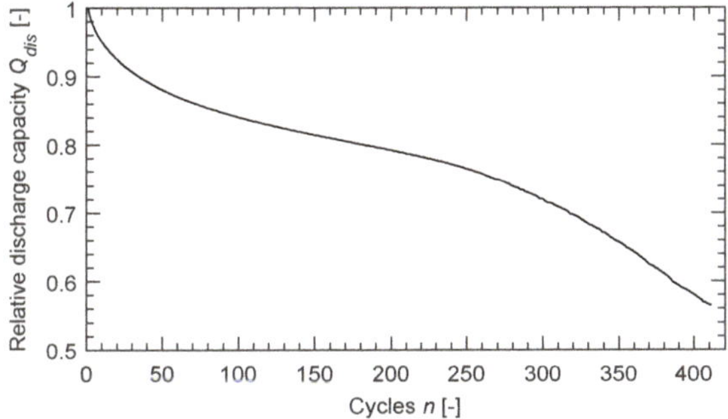

Figure 3. Relative discharge capacity of the simulated cell over the life time. A linear ageing phase following by a non-linear aging phase are observable.

Looking at the equivalent thickness of the lithium plating layer on the surface of anode particles in Figure 4A, we can see that from the 116th cycle Li-P started at the separator side of the negative electrode and the layer thickness increased by continuing the cycling of the cell. During the whole 400 simulated cycles, no Li-plating occurred at the current collector side of the anode. The total surface layer, Li-P, and SEI, together with equivalent thickness of each layer is shown in Figure 4B. From cycle 116 until 230, Li-plating showed a more moderate increase rate in comparison to cycles after 230 which is when the cell began nonlinear aging behavior. In contrast to Li-P, the SEI layer grew with a high rate during the first 50 cycles and after that increased more moderately. The decrease in the SEI layer's growth rate is due to the limited electrolyte diffusion through the formed layer as well as the lower EC concentration in the electrolyte as it became consumed through the SEI formation reaction. In contrast to lithium plating, which depends on the location along the anode thickness, SEI layer growth was uniform across the anode. The SEI and Li-P at the separator, grew to around $d_{SEI} \approx 800$ nm and $d_{Li-P} \approx 120$ nm. Kindermann et al. [38] simulated the SEI layer with $d_{SEI} \approx 600$ nm. Separately Petzl et al. [39] in their experimental low-temperature study measured a $d_{Li-P} \approx 5$ μm after 120 cycles at $-22\,°$C with 1C.

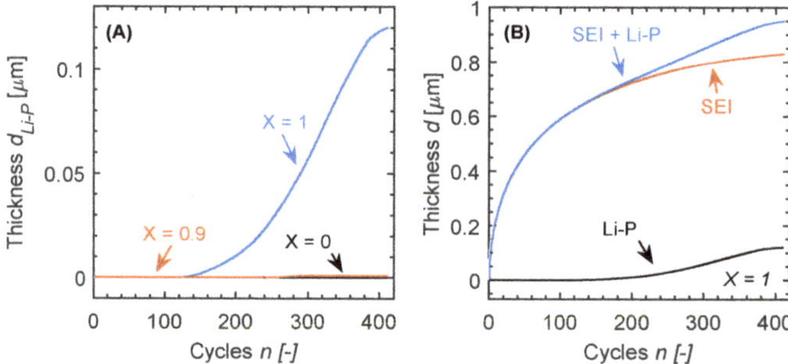

Figure 4. (**A**) Thickness of the plated lithium layer on the surface of negative electrode particles at the current collector (X = 0), 0.9 of the relative anode thickness (X = 0.9), and the separator side (X = 1). Over the cycle numbers, no lithium plating happened at the current collector side but it increased moving toward the separator side. (**B**) Total surface layer thickness as well as equivalent thicknesses of plated lithium and SEI (solid electrolyte interface) layers separately. Since the cell has the maximum amount lithium plating at the separator side, only the layers at X = 1 are displayed.

4. Discussion

To study and track the reasons leading to lithium plating, starting from cycle 116 in our cell, we plotted the ionic conductivity of solid particles, the electrolyte, and electrode/separator interface over the constant current (CC) charge period for the anode and cathode, as shown in Figure 5 and 6 respectively. Figure 5A shows calculated ionic conductivity for region A which are the solid active particles of the negative electrode from start until end of the CC charge period for the 116th cycle. Two points of X = 1 and X = 0.9 are chosen to be shown. X = 1 for the anode means the separator side. X = 0.9 is also displayed as it is the furthest point of the anode from the separator that shows lithium deposition over the whole simulated cycle life. As expected, the ionic conductivity of the particles declined while the SOC increased. At the end of the charge, particles closer to the separator side were at a higher SOC and therefore showed poorer ionic conductivity in comparison to particles with a higher distance to the separator. The beginning of Li-plating is where a short plateau is observable in the conductivity trend at the end of CC charge. Effective electrolyte ionic conductivity in Figure 5B is showing a similar trend. It displayed a higher value during early charge stages in comparison to the end of the charge as well as poorer particle ionic conductivity closer to the separator. This is explainable as the side reactions and surface layer formation was happening more at the separator side, leading to more porosity reduction which is equal to less electrolyte volume fraction. The last transport region of the anode that we include in this study is the electrolyte/solid particles interface. As shown in Figure 5C, the ionic conductivity increased while charging until a local SOC of 50%. By continuing the CC charge process we can observe a decline in conductivity values followed by a plateau which is the start of lithium plating. Comparing $\sigma_{s,a}$ to the corresponding Figure 6A, shows that anode active particles are having a better ionic conductivity, except for the end of the charge process which were slightly smaller than the cathode particle conductivity values. Figure 6B shows that electrolyte had a better effective ionic conductivity of factor 8 at the cathode side. It is the same when we compare their interface conductivity as well. Figure 6C shows that $\sigma_{i,c}$ is one order of magnitude bigger than $\sigma_{i,a}$.

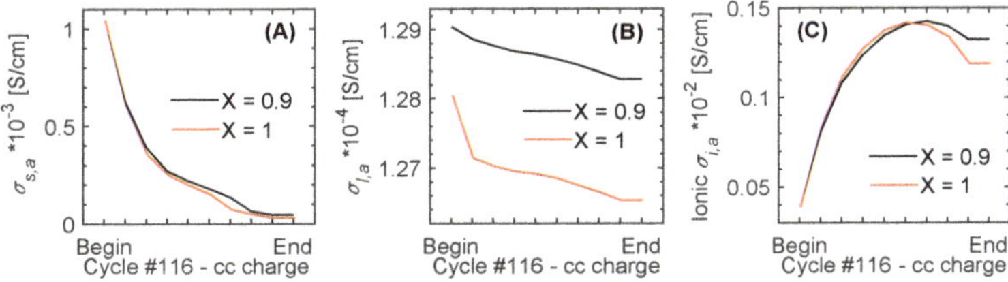

Figure 5. Ionic conductivity of the simulated cell at cycle 116 for the negative electrode side. Start of Li-plating is when a plateau is formed. (**A**) Is the calculated ionic conductivity for region A over start until end of CC (current) charge period of the 116th cycle. X = 1 is at the separator side and X = 0.9 is at 0.9 of anode thickness closer to the separator. (**B**) is the effective ionic conductivity of electrolyte in the negative electrode over start until end of CC charge period of the 116th cycle. Data comes from the model. (**C**) Is the calculated ionic conductivity for region B over start until end of the CC charge period of the 116th cycle.

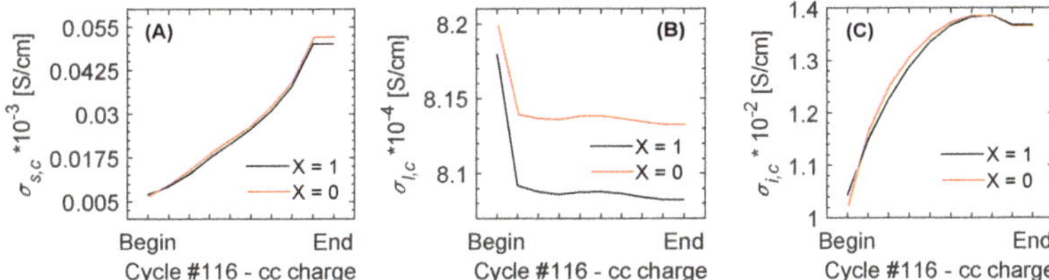

Figure 6. Ionic conductivity of the simulated cell at cycle 116 for the positive electrode side. (**A**) Is the calculated ionic conductivity for region E over start until end of the CC charge period of the 116th cycle. X = 0 is at separator side and X = 1 is at the current collector of the cathode. (**B**) Is the effective ionic conductivity of the electrolyte in the positive electrode over start until end of the CC charge period of the 116th cycle. Data comes from the model. (**C**) Is the calculated ionic conductivity for region D over start until end of the CC charge period of the 116th cycle.

For comparison, the same is plotted in Figures 7 and 8 for the ionic conductivity results of the cell at cycle 10 where the cell is not aged and shows no trace of lithium plating. Looking at Figures 7A and 8A, we can see that values neither for the cathode nor the anode change significantly. In comparison to cycle 10, the final value at the end of charge of cycle 116 for both cathode and anode at the separator side declined. For the positive electrode, the cathode particles could not de-intercalate fully during the 10th cycle. To explain the anode behavior, listed data in Table 3 is helpful. Considering the $\sigma_{s,a}$ at different cycles, it is observable that by aging, the anode distribution of lithium ions becomes deficient so that by increasing the cycle number $\sigma_{s,a}$ at the current collector side shows higher values. This means by increasing the cycle number, during a charge process, particles at X = 0 only charged to the lower SOCs. The reverse is observable for the particles at the separator side of the anode. This can lead the cell to favorable conditions for Li deposition. Figures 7B and 8B show that although electrolyte conductivity does not show a significant change at the cathode, the anode values declined by factor of 4. A similar behavior is observable for the anode interface ionic conductivity. Figure 7C shows that conductivity values of the 10th cycle are about 4 times bigger than those at cycle 116. However no significant variation is shown in Figure 8C in comparison to cycle 116. These two behaviors are another factor that cause Li-plating.

Table 3. Ionic conductivity for solid particles of active materials σ_s, electrolyte in the porous electrodes σ_l, and electrode/electrolyte interface σ_i at cycle number 10, 50, 100, 116, 150, and 230. Cycle 116 is the start of Li-P and cycle 230 is the start of nonlinear aging behavior of the cell. σ_s and σ_i are calculated from Nernst–Einstein relation. σ_l is calculated directly through the model.

Cycle	Charge – CC	σ_s [S/cm]	σ_l [S/cm]	σ_i [S/cm]
		An.	**An.**	**An.**
10	Begin	1.0×10^{-3}	$(4.8\text{-}4.7) \times 10^{-4}$	1.3×10^{-3}
	End	$(7.3\text{-}3.5) \times 10^{-5}$	$(4.7\text{-}4.6) \times 10^{-4}$	$(4.5\text{-}4.0) \times 10^{-3}$
50	Begin	$(9.9\text{-}10.3) \times 10^{-4}$	$(2.5\text{-}2.4) \times 10^{-4}$	0.7×10^{-3}
	End	$(14.7\text{-}3.7) \times 10^{-5}$	$(2.5\text{-}2.4) \times 10^{-4}$	$(2.6\text{-}2.2) \times 10^{-3}$
100	Begin	$(9.2\text{-}10.4) \times 10^{-4}$	1.5×10^{-4}	$(0.5\text{-}0.4) \times 10^{-3}$
	End	$(18.3\text{-}3.3) \times 10^{-5}$	$(1.5\text{-}1.4) \times 10^{-4}$	$(1.6\text{-}1.4) \times 10^{-3}$
116	Begin	$(9.1\text{-}10.4) \times 10^{-4}$	1.3×10^{-4}	$(0.5\text{-}0.4) \times 10^{-3}$
	End	$(19.1\text{-}3.1) \times 10^{-5}$	1.3×10^{-4}	$(1.4\text{-}1.2) \times 10^{-3}$
150	Begin	$(8.8\text{-}10.5) \times 10^{-4}$	1.0×10^{-4}	$(0.4\text{-}0.3) \times 10^{-3}$
	End	$(20.1\text{-}2.8) \times 10^{-5}$	$(1.1\text{-}1.0) \times 10^{-4}$	$(1.1\text{-}0.9) \times 10^{-3}$
230	Begin	$(7.2\text{-}10.5) \times 10^{-4}$	$(0.6\text{-}0.5) \times 10^{-4}$	$(0.3\text{-}0.2) \times 10^{-3}$
	End	$(21.7\text{-}0.2) \times 10^{-5}$	$(0.6\text{-}0.5) \times 10^{-4}$	$(0.7\text{-}0.3) \times 10^{-3}$
		Ca.	**Ca.**	**Ca.**
10	Begin	$(3.4\text{-}4.0) \times 10^{-6}$	9.2×10^{-4}	0.8×10^{-2}
	End	$(5.8\text{-}5.7) \times 10^{-5}$	9.2×10^{-4}	1.3×10^{-2}
50	Begin	$(5.1\text{-}5.7) \times 10^{-6}$	$(8.8\text{-}8.7) \times 10^{-4}$	$(0.9\text{-}1.0) \times 10^{-2}$
	End	$(5.6\text{-}5.4) \times 10^{-5}$	8.7×10^{-4}	1.4×10^{-2}
100	Begin	$(6.2\text{-}6.7) \times 10^{-6}$	8.3×10^{-4}	1.0×10^{-2}
	End	$(5.3\text{-}5.1) \times 10^{-5}$	$(8.3\text{-}8.2) \times 10^{-4}$	1.4×10^{-2}
116	Begin	$(6.5\text{-}7.0) \times 10^{-6}$	8.2×10^{-4}	1×10^{-2}
	End	$(5.2\text{-}5.0) \times 10^{-5}$	8.1×10^{-4}	1.4×10^{-2}
150	Begin	$(6.9\text{-}7.5) \times 10^{-6}$	8×10^{-4}	$(1.0\text{-}1.1) \times 10^{-2}$
	End	$(5.0\text{-}4.7) \times 10^{-5}$	$(7.9\text{-}7.8) \times 10^{-4}$	1.4×10^{-2}
230	Begin	$(8.1\text{-}9.2) \times 10^{-6}$	7.6×10^{-4}	1.1×10^{-2}
	End	$(5.7\text{-}5.6) \times 10^{-5}$	7.4×10^{-4}	1.3×10^{-2}

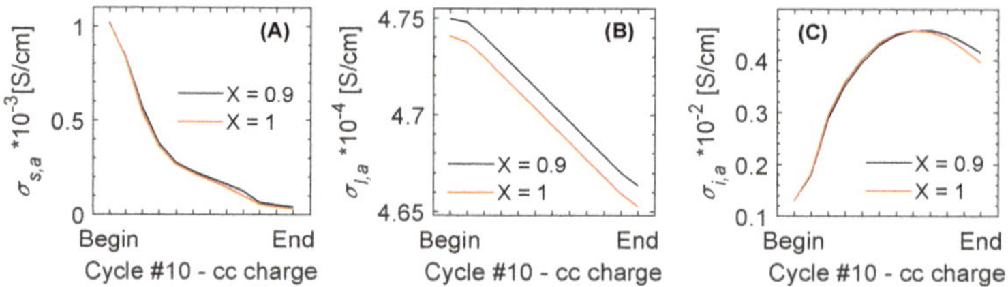

Figure 7. Ionic conductivity of the simulated cell at cycle 10 for the negative electrode side. (**A**) Is the calculated ionic conductivity for region A over start until end of the CC charge period of the 10th cycle. X = 1 is at the separator side and X = 0.9 is at 0.9 of relative anode thickness closer to the separator. (**B**) Is the effective ionic conductivity of the electrolyte in the negative electrode over start until end of the CC charge period of the 10th cycle. Data comes from the model. (**C**) Is the calculated ionic conductivity for region B over start until the end of the CC charge period of the 10th cycle.

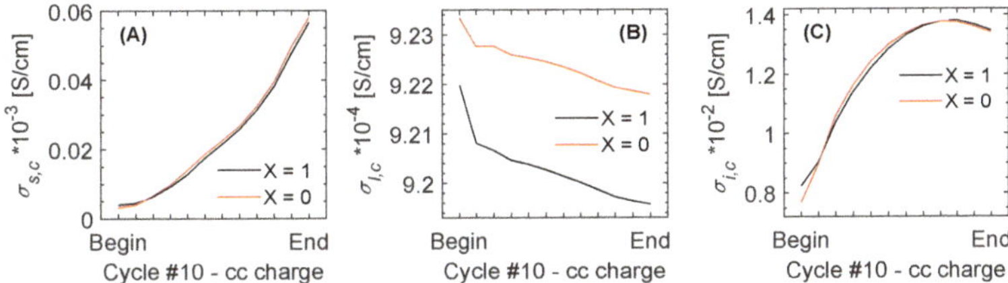

Figure 8. Ionic conductivity of the simulated cell at cycle 10 for the positive electrode side. (**A**) Is the calculated ionic conductivity for region E over start until end of the CC charge period of the 10th cycle. X = 0 is at the separator side and X = 1 is at the current collector of the cathode. (**B**) Is the effective ionic conductivity of electrolyte in the positive electrode over start until end of the CC charge period of 10th cycle. Data comes from the model. (**C**) Is the calculated ionic conductivity for region D over start until end of the CC charge period of the 10th cycle.

5. Conclusions

In this work, we studied the aging behavior of a Li-ion cell and investigated the ionic conductivity variation of the cell and its relation to the appearance of unwanted lithium deposition on the surface of negative electrode particles from the 116th cycle. We introduced five regions that explained the transport mechanisms of lithium from one electrode to the other, including the Li^+ charge transfer, Li solid diffusion, and Li^+ diffusion through the electrolyte. Simulation data and ionic conductivity calculations showed that both interface and electrolyte conductivity of the anode ($\sigma_{i,a}$ and $\sigma_{l,a}$) at cycle 116 were four times smaller than the not-aged values. Moreover solid particle conductivity of the anode $\sigma_{s,a}$ showed that the poor distribution of Li in the anode particles along the x direction led to local overcharging on the separator side. These effects together caused the cell to plate lithium from the 116th cycle.

Author Contributions: Conceptualization, S.M.B. and K.P.B.; methodology, S.M.B. and K.P.B.; software, S.M.B.; formal analysis, S.M.B.; writing–original draft, S.M.B. and K.P.B. ; writing–review and editing, S.M.B.; visualization, S.M.B.; supervision, K.P.B.; funding acquisition, K.P.B.

Funding: This research was funded by Robert Bosch GmbH

Acknowledgments: The authors thank Robert Bosch GmbH for their financial support through the Bosch Promotionskolleg. The author would also like to thank D. Müller for their constructive criticism of the manuscript.

Nomenclature

Abbreviations

P2D	pseudo two-dimensional
SEI	solid electrolyte interface
Li-P	lithium plating
Li-ion	lithium ion
CTP	charge transfer process
SPD	solid diffusion process
SOC	state of charge
EC	ethylene carbonate
CC	constant current
CV	constant voltage
An.	anode
Ca.	cathode

Lit. literature
Cal. calculated

Symbols

j ionic current density
j_0 exchange current density
v drift velocity
u electrical mobility of ions
Z valence number
q charge
K Boltzamn constant
R ideal gas constant
F Faraday constant
D diffusion coefficient
ϕ potential
i current density
σ conductivity
ζ electric field
α transfer coefficient
η surface overpotential
E potential
T temperature
C concentration
T temperature
ϵ volume fraction
γ Bruggeman exponent
t characteristic time
L thickness
L_s separator thickness

Subscripts and Superscripts

ion ion species
tot total
int intercalation
SEI solid electrolyte interface
LiP lithium plating
a anodic, anode
c cathodic, cathode
s solid
a,c,s anode, cathode, separator
l liquid
i interface
eq equilibrium
eff effective
max maximum
ave average
ref reference

References

1. Brandt, K. Historical development of secondary lithium batteries. *Solid State Ionics* **1994**, *69*, 173–183. [CrossRef]
2. Besenhard, J. The electrochemical preparation and properties of ionic alkali metal-and NR4-graphite intercalation compounds in organic electrolytes. *Carbon* **1976**, *14*, 111–115. [CrossRef]
3. Whittingham, M.S. Electrical energy storage and intercalation chemistry. *Science* **1976**, *192*, 1126–1127. [CrossRef] [PubMed]
4. Slane, S.M.; Foster, D.L. Lithium Ion Rechargeable Intercallation Cell. US Patent No. 07/625,181, 7 July 1992.

5. Yoshino, A. Development of the lithium-ion battery and recent technological trends. In *Lithium-Ion Batteries*; Elsevier: Amsterdam, The Netherlands, 2014; pp. 1–20.

6. Almeida, A.; Sousa, N.; Coutinho-Rodrigues, J. Quest for Sustainability: Life-Cycle Emissions Assessment of Electric Vehicles Considering Newer Li-Ion Batteries. *Sustainability* **2019**, *11*, 2366. [CrossRef]

7. Onat, N.; Kucukvar, M.; Tatari, O. Towards life cycle sustainability assessment of alternative passenger vehicles. *Sustainability* **2014**, *6*, 9305–9342. [CrossRef]

8. Burns, J.; Stevens, D.; Dahn, J. In-situ detection of lithium plating using high precision coulometry. *J. Electrochem. Soc.* **2015**, *162*, A959–A964. [CrossRef]

9. Fleischhammer, M.; Waldmann, T.; Bisle, G.; Hogg, B.I.; Wohlfahrt-Mehrens, M. Interaction of cyclic ageing at high-rate and low temperatures and safety in lithium-ion batteries. *J. Power Sources* **2015**, *274*, 432–439. [CrossRef]

10. Waldmann, T.; Hogg, B.I.; Wohlfahrt-Mehrens, M. Li plating as unwanted side reaction in commercial Li-ion cells—A review. *J. Power Sources* **2018**, *384*, 107–124. [CrossRef]

11. Uhlmann, C.; Illig, J.; Ender, M.; Schuster, R.; Ivers-Tiffée, E. In situ detection of lithium metal plating on graphite in experimental cells. *J. Power Sources* **2015**, *279*, 428–438. [CrossRef]

12. Doh, C.H.; Han, B.C.; Jin, B.S.; Gu, H.B. Structures and formation energies of LixC6 (x = 1-3) and its homologues for lithium rechargeable batteries. *Bull. Korean Chem. Soc* **2011**, *32*, 2045–2050. [CrossRef]

13. Legrand, N.; Knosp, B.; Desprez, P.; Lapicque, F.; Raël, S. Physical characterization of the charging process of a Li-ion battery and prediction of Li plating by electrochemical modelling. *J. Power Sources* **2014**, *245*, 208–216. [CrossRef]

14. Atkins, P.W.; De Paula, J.; Keeler, J. *Atkins' Physical Chemistry*; Oxford University Press: Oxford, UK, 2018.

15. Hein, S.; Latz, A. Influence of local lithium metal deposition in 3D microstructures on local and global behavior of Lithium-ion batteries. *Electrochim. Acta* **2016**, *201*, 354–365. [CrossRef]

16. Wu, M.S.; Chiang, P.C.J.; Lin, J.C. Electrochemical investigations on advanced lithium-ion batteries by three-electrode measurements. *J. Electrochem. Soc.* **2005**, *152*, A47–A52. [CrossRef]

17. Lin, H.P.; Chua, D.; Salomon, M.; Shiao, H.; Hendrickson, M.; Plichta, E.; Slane, S. Low-temperature behavior of Li-ion cells. *Electrochem. Solid-State Lett.* **2001**, *4*, A71–A73. [CrossRef]

18. Waldmann, T.; Hogg, B.I.; Kasper, M.; Grolleau, S.; Couceiro, C.G.; Trad, K.; Matadi, B.P.; Wohlfahrt-Mehrens, M. Interplay of operational parameters on lithium deposition in lithium-ion cells: systematic measurements with reconstructed 3-electrode pouch full cells. *J. Electrochem. Soc.* **2016**, *163*, A1232–A1238. [CrossRef]

19. Persson, K.; Sethuraman, V.A.; Hardwick, L.J.; Hinuma, Y.; Meng, Y.S.; Van Der Ven, A.; Srinivasan, V.; Kostecki, R.; Ceder, G. Lithium diffusion in graphitic carbon. *J. Phys. Chem. Lett.* **2010**, *1*, 1176–1180. [CrossRef]

20. Persson, K.; Hinuma, Y.; Meng, Y.S.; Van der Ven, A.; Ceder, G. Thermodynamic and kinetic properties of the Li-graphite system from first-principles calculations. *Phys. Rev. B* **2010**, *82*, 125416. [CrossRef]

21. Smart, M.; Ratnakumar, B. Effects of electrolyte composition on lithium plating in lithium-ion cells. *J. Electrochem. Soc.* **2011**, *158*, A379–A389. [CrossRef]

22. Ishihara, Y.; Miyazaki, K.; Fukutsuka, T.; Abe, T. Kinetics of lithium-ion transfer at the interface between Li4Ti5O12 thin films and organic electrolytes. *ECS Electrochem. Lett.* **2014**, *3*, A83–A86. [CrossRef]

23. Abe, T.; Sagane, F.; Ohtsuka, M.; Iriyama, Y.; Ogumi, Z. Lithium-ion transfer at the interface between lithium-ion conductive ceramic electrolyte and liquid electrolyte—A key to enhancing the rate capability of lithium-ion batteries. *J. Electrochem. Soc.* **2005**, *152*, A2151–A2154. [CrossRef]

24. Jow, T.R.; Delp, S.A.; Allen, J.L.; Jones, J.P.; Smart, M.C. Factors Limiting Li$^+$ Charge Transfer Kinetics in Li-Ion Batteries. *J. Electrochem. Soc.* **2018**, *165*, A361–A367. [CrossRef]

25. Arora, P.; Doyle, M.; White, R.E. Mathematical modeling of the lithium deposition overcharge reaction in lithium-ion batteries using carbon-based negative electrodes. *J. Electrochem. Soc.* **1999**, *146*, 3543–3553. [CrossRef]

26. Tang, M.; Albertus, P.; Newman, J. Two-dimensional modeling of lithium deposition during cell charging. *J. Electrochem. Soc.* **2009**, *156*, A390–A399. [CrossRef]

27. Jiang, F.; Peng, P. Elucidating the performance limitations of lithium-ion batteries due to species and charge transport through five characteristic parameters. *Sci. Rep.* **2016**, *6*, 32639. [CrossRef] [PubMed]

28. Doyle, M.; Fuller, T.F.; Newman, J. Modeling of galvanostatic charge and discharge of the lithium/polymer/insertion cell. *J. Electrochem. Soc.* **1993**, *140*, 1526–1533. [CrossRef]

29. Fuller, T.F.; Doyle, M.; Newman, J. Simulation and optimization of the dual lithium ion insertion cell. *J. Electrochem. Soc.* **1994**, *141*, 1–10. [CrossRef]

30. Doyle, C.M. Design and Simulation of Lithium Rechargeable Batteries. Ph.D. Thesis, University of California, Berkeley, CA, USA, 1995.

31. Doyle, M.; Newman, J.; Gozdz, A.S.; Schmutz, C.N.; Tarascon, J.M. Comparison of modeling predictions with experimental data from plastic lithium ion cells. *J. Electrochem. Soc.* **1996**, *143*, 1890–1903. [CrossRef]

32. Müller, D.; Dufaux, T.; Birke, K.P. Model-Based Investigation of Porosity Profiles in Graphite Anodes Regarding Sudden-Death and Second-Life of Lithium Ion Cells. *Batteries* **2019**, *5*, 49. [CrossRef]

33. Yang, X.G.; Leng, Y.; Zhang, G.; Ge, S.; Wang, C.Y. Modeling of lithium plating induced aging of lithium-ion batteries: Transition from linear to nonlinear aging. *J. Power Sources* **2017**, *360*, 28–40. [CrossRef]

34. Darling, R.; Newman, J. Modeling side reactions in composite li y mn2 o 4 electrodes. *J. Electrochem. Soc.* **1998**, *145*, 990–998. [CrossRef]

35. Cabañero, M.A.; Altmann, J.; Gold, L.; Boaretto, N.; Müller, J.; Hein, S.; Zausch, J.; Kallo, J.; Latz, A. Investigation of the temperature dependence of lithium plating onset conditions in commercial Li-ion batteries. *Energy* **2019**, *171*, 1217–1228. [CrossRef]

36. Eddahech, A.; Briat, O.; Vinassa, J.M. Thermal characterization of a high-power lithium-ion battery: Potentiometric and calorimetric measurement of entropy changes. *Energy* **2013**, *61*, 432–439. [CrossRef]

37. Safari, M.; Morcrette, M.; Teyssot, A.; Delacourt, C. Multimodal physics-based aging model for life prediction of Li-ion batteries. *J. Electrochem. Soc.* **2009**, *156*, A145–A153. [CrossRef]

38. Kindermann, F.M.; Keil, J.; Frank, A.; Jossen, A. A SEI modeling approach distinguishing between capacity and power fade. *J. Electrochem. Soc.* **2017**, *164*, E287–E294. [CrossRef]

39. Petzl, M.; Kasper, M.; Danzer, M.A. Lithium plating in a commercial lithium-ion battery—A low-temperature aging study. *J. Power Sources* **2015**, *275*, 799–807. [CrossRef]

Article

Post-Mortem Analysis of Inhomogeneous Induced Pressure on Commercial Lithium-Ion Pouch Cells and Their Effects

Georg Fuchs [1,2], **Lisa Willenberg** [1,2], **Florian Ringbeck** [1,2] **and Dirk Uwe Sauer** [1,2,3,4,*]

1   Chair of Electrochemical Energy Conversion and Storage Systems, Institute for Power Electronics and Electrical Drives (ISEA), RWTH Aachen University, Jaegerstr. 17/19, 52066 Aachen, Germany; Georg.Fuchs@isea.rwth-aachen.de (G.F.); lisa.willenberg@isea.rwth-aachen.de (L.W.); florian.ringbeck@isea.rwth-aachen.de (F.R.)
2   Jülich Aachen Research Alliance, JARA-Energy, Forschungszentrum Jülich GmbH, 52425 Jülich, Germany
3   Institute for Power Generation and Storage Systems (PGS), E.ON ERC, RWTH Aachen University, Mathieustraße 10, 52074 Aachen, Germany
4   Helmholtz Institute Münster (HI MS), IEK-12, Forschungszentrum Jülich GmbH, 52425 Jülich, Germany
*   Correspondence: batteries@isea.rwth-aachen.de

Received: 1 November 2019; Accepted: 21 November 2019; Published: 27 November 2019

Abstract: This work conducts a post-mortem analysis of a cycled commercial lithium-ion pouch cell under an induced inhomogeneous pressure by using a stainless-steel sphere as a force transmitter to induce an inhomogeneous pressure distribution on a cycled lithium-ion battery. After the cycling, a macroscopic and microscopic optical analysis of the active and passive materials was executed. Also, scanning electron microscopy was used to analyze active material particles. The sphere shape results in a heterogenic pressure distribution on the lithium-ion battery and induces a ring of locally high electrochemical activity, which leads to lithium plating. Furthermore, a surface layer found on the anode, which is a possible cause of electrolyte degradation at the particle–electrolyte interface. Significant deformation and destruction of particles by the local pressure was observed on the cathode. The analysis results validate previous simulations and theories regarding lithium plating on edge effects. These results show that pressure has a strong influence on electrolyte-soaked active materials.

Keywords: inhomogeneous pressure; localized plating; mechanical stress; separator; tortuosity; ICP-OES; scanning electron microscope; lithium-ion battery

1. Introduction

Today, lithium-ion batteries (LIBs) are omnipresent in everyday life. After their commercial introduction, LIBs have largely superseded other battery technologies due to their superior properties. They are used in nearly all portable devices such as mobile phones, smartwatches, cameras, laptops, e-cigarettes, and power tools. Furthermore, LIBs are increasingly penetrating the automotive market, where the battery is a crucial component for mobility. The battery cells of an electric car still account for the largest share of vehicle costs, and therefore, the lifetime and reliability of this component is essential. In almost every application, LIBs are integrated into housings and must be mechanically clamped. Previous work indicates that the design of a battery module consisting of several battery cells in terms of mechanical bracing is the crucial element of lifetime and reliability. Cannarella et al. showed that too much initial pressure applied to LIBs decreases their lifetime due to closed pores in the separator, which causes undesirable degradation effects [1].

Further work by the same research group shows that these effects can include local lithium deposition and dendrite growth, affecting electrochemical kinetics [2–5]. Wünsch et al. show that not only the initial pressure is an essential factor for the lifetime of LIBs, but also the type of bracing.

Usually, a rigid bracing is used in the construction of LIB modules. Since LIBs with a graphite anode expand during the charging processes, under thermal stress and over a lifetime, the pressure on the cells in the module increases over time [6,7]. According to Wünsch et al., the cycle life can be significantly increased with a quasi-constant force bracing in contrast to the usually applied fixed bracing [7]. From this, it can be deduced that the application of constant pressure can significantly improve the lifetime of LIB modules.

While the work discussed in the previous paragraph analyzed the influence of homogenous pressure in LIB, also inhomogeneous pressure must be considered for real-life applications as the mechanical bracing used in battery modules does not always ensure that the pressure homogeneously distributed. The work of Cannarella and Liu shows that local mechanical deformations of the separator can cause local lithium plating due to inhomogeneous ion flux distributions [4,5]. Furthermore, the work of Tang et al. has shown that lithium plating occurs preferentially at electrodes edges of LIBs due to geometrics effects, which generates overpotential at the edges of the electrodes and leads to conditions which favor lithium plating [8]. Other investigations by Rahe et al. using Nano X-Ray tomography show particle cracks and current-collector corrosion on the cathode side [9], which also leads to a local pressure increase and strongly influences the porosity and thus the tortuosity of the active, cathode and anode, as well as the passive, separator, materials. Another effect observed by Waldmann et al. is that mechanical deformations occur at the jelly roll of a round cell at the end of lifetime. This mechanical deformation induces local inhomogeneous pressure on the active materials [10], which leads to inhomogeneous ion flux distributions and causes lithium plating. Bach et al. linked the sudden degradation effect at the end-of-life of round cells triggered by the appearance of lithium plating confined to small characteristic areas at the jelly roll, generated by heterogeneous compression, and mention the importance of cell and pack design considering a well mechanical bracing without inducing unwanted effects [11].

As presented in the previous paragraph, mechanical defects of the separator are linked to lithium plating. For lithium plating to occur, the local anode potential must be below 0 V vs. Li/Li+ [8,12]. Only under this condition, it is energetically more favorable for the lithium-ions to bond with each other and form metallic lithium. Especially for graphite anodes, lithium dendrites will continue to grow under this condition. Furthermore, as the lithium metal is exposed to the electrolyte, electrolyte degradation products (EDPs) will be produced, so that the lithium dendrites are covered with a thick lithium–electrolyte interface (LEI) [13]. This degradation process directly leads to capacity loss and gas formation. Also, localized defects can lead to local lithium deposition and dendrite growth, and the plating of lithium could trigger a short circuit by the penetration of a thin separator. In the worst case, dendrites can lead to catastrophic failure due to associate these internal short circuits and lead the LIB into the thermal runaway, as it happens in case of the Samsung Galaxy Note 7 [12,14].

Most of the previous work concerning local deformation of the separator and the resulting lithium plating was only carried out on half-cell experiments in the laboratory. The separator was locally deformed before installation in the half-cell causing locally closed separator pores, which lead to uneven ion currents [2,4,5]. There is no work available that has investigated local induced inhomogeneities in commercial pouch cells. Therefore, in this work, local pressure is applied to a commercial high energy pouch cell manufactured by Kokam SLPB526495, and a local deformation applied to the LIB. A stainless-steel sphere used as a force transmitter, which leads to radial inhomogeneous pressure distribution. The LIB cycled under this mechanical load. A post-mortem analysis (PMA) of the cell stack and the active and passive materials was then performed to analyze inhomogeneities in state of charge, surface layer formation, and particle morphology.

2. Materials and Methods

2.1. Investigated Lithium-Ion Battery

For the experiment, a LIB of the manufacturer Kokam SLPB526495 with a capacity of 3.3 Ah is used. The LIB is specified for a charging current of 2C and a voltage range of 2.7 V to 4.2 V between 0 to 45 °C. According to the manufacturer, the cell is made of a graphite anode, a Li(NiCo)O$_2$ cathode, and an EC/EMC mixture with LiPFO$_6$ as an electrolyte. Figure 1 shows the mechanical structure of the LIB which consists of two electrode stacks connected in parallel. The first stack consists of six double-sided coated anodes, five double-sided coated cathodes, and one one-sided coated cathode. The second stack consists of one additional double-sided cathode. Each stack individually wrapped in a Z-shaped separator.

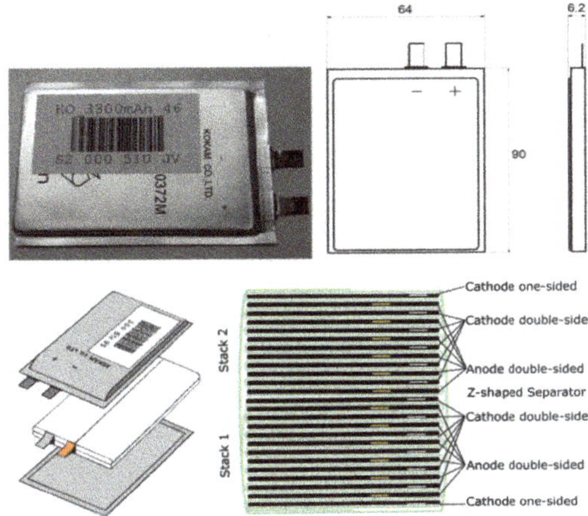

Figure 1. Dimensions (in mm) and internal construction of the Kokam SLP526495 cell.

2.2. Experimental Setup

For examination and cycling of the lithium-ion battery, the cell test system Digatron MCFT 20-05-50ME with a current measurement precision of 0.2% is used. The measurement setup during cell cycling depicted in Figure 2a. In order to exert inhomogenous pressure, a stainless-steel sphere with a diameter of 4 mm placed on the center of the cell surface. A quasi-statically force is applied to the sphere via a hydraulic press so that the cell deforms—the applied force is measured with a load cell placed between the sphere and the hydraulic press. The applied pressure was chosen based on the analysis of Wang et al. and Tran et al. to apply as much force as possible without damaging the active material particles. Wang et al. suggests a pressure of 10 to 20 MPa as optimal pressure regarding the compression and contacting of graphite particles during the calendaring process of anodes after coating [15]. According to Tran et al., the pressure during calendaring of NCA cathodes is significantly higher and optimal at 500 to 692 MPa [16] . Therefore, a pressure of 20 MPa was chosen for the experiments.

The applied force by the press in order exert this pressure is calculate based on the formula for Hertzian pressure for a sphere-plane configuration. For this calculation, the Young's modulus of the LIB must be known. As this quantity varies with cell type and not stated in the cell's datasheet, different forces were applied to the cell, and the penetration depth was measured using a dial gauge. After this series of tests, a force of 1.4 kN with a penetration depth of 0.8 mm was determined to achieve a maximum pressure of 20 MPa in the center of the area under the sphere. Assuming the LIB to be an elastic body, Young's modulus can be approximated to be 210 MPa. Nevertheless, since the active materials in LIB are porous and consist of multi-particle components, the force will be distributed over the particles. Therefore, the maximum local pressure could be much higher due to the tiny contact areas of the complex porous particle system. By applying the pressure, defects induced in the materials beneath the sphere. As the pressure decreases towards the edge, Figure 2b–d shows a transition area created where the porosity increases and therefore creates an unevenly increasing ion flux.

Another essential effect of particles in LIBs is the swelling during the lithiation process, and this adds additional local pressure on the contact areas. Also, the deposition of lithium metal will lead to an increase in the local pressure due to thickness change and stress the local particles furthermore.

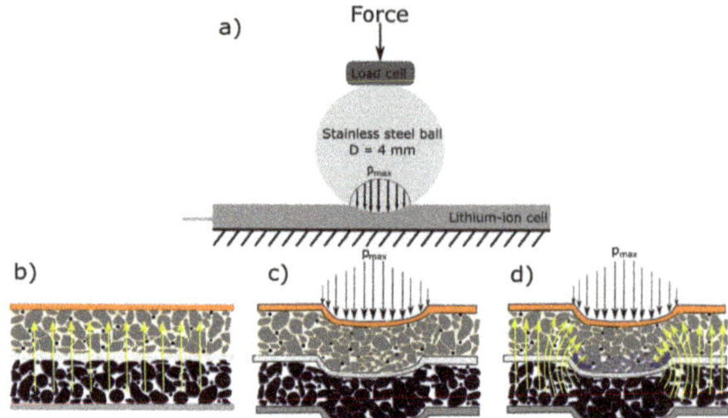

Figure 2. Setup and principle of the experiment (**a**) Li-cell with stainless-steel sphere and load cell (**b**) non-deformed state (homogeneous ionic current distribution yellow arrows) (**c**) deformed state due to inhomogeneous induced pressure distribution (**d**) deformed state with inhomogeneous flow field of ionic current density.

To analyze the cell aging under the above-described inhomogeneous mechanical pressure, the tested cell then cycled several times. Before cycling, the pressure adjusted to 1.3 kN at a cell voltage of 3.55 V and a temperature of 20 °C. After the application of the pressure, the cell relaxed for 30 min and was then charged with 1C to 4.15 V until the force increased to 1.4 kN. The force increases due to the swelling of the graphite anode particles during the charging process [17]. The LIB has then cycled in CC/CV mode three times with a current of 1C between 4.2 V and 2.7 V. The CV-phase takes 30 min in each cycle. During cycling under the hydraulic press, the force fluctuated by a maximum of 150 N. Finally, the cell was fully charged with a current of 1C to 4.2 V and then opened for a post-mortem analysis.

To evaluate the influence of inhomogeneous pressure during the operation of a commercial, two stacked pouch, cell one cycled, and one uncycled were analyzed in this experiment. Subsequently, the surface of carefully selected areas analyzed by scanning electron microscopy analysis (SEM). The electron microscope used was a Leo Supra 35 VP from Carl Zeiss AG with an INCA Energy 200 EDS detector from Oxford Instruments. Best images were recorded using an in-lens BSE detector in a working distance of 7 mm at a high vacuum (10 to 6 mbar) with an acceleration voltage of 5 kV. EDS spot measurements or mappings were carried out at 10 kV. The uncycled cell was opened in a pristine state and acted as a reference cell. Individual dry layers of the anode and cathodes were pressurized with a homogeneous pressure of 30 MPa to have a comparison to the inhomogeneous pressure effects on the electrodes of the LIB.

The cycled LIB was opened fully charged at 4.2 V under argon atmosphere in a glove box, as in the fully charged state, inhomogeneities within the cell can be better identified. A particular property of graphite is a color change depending on the lithiation, i.e., the state of charge. While completely delithiated graphite that corresponds to an SOC of 0% is black or gray and between a lithiation state 0% to 30% the color is very similar to that of the pristine graphite (gray-black), it discolors with increasing lithiation over dark blue (30 to 50% SOC) and red (50 to 90% SOC) to golden (90 to 100%

SOC) [18,19]. This property makes it easy to determine the lithiation and thus SOC inhomogeneities of the opened cell.

The cathode electrode compositions are measured with Varian 725 induced coupled plasma-optical emission spectrometer (ICP-OES) (Agilent, Santa Clare, United States of America). For this, a disc of the double-coated electrode with a diameter of 20 mm is taken. This sample is washed with Dimethylcarbonat (Dimethylcarbonat Msynth plus, Merck KGaA, Darmstadt, Germany) before it is dissolved in aqua regia. The solution was filled with distilled water until a 100 ml solution was obtained. This solution was analyzed with the ICP-OES.

3. Results

3.1. Cathode Compositions

The ICP-OES and EDX analysis shows that the cathode of the cell contains 64% Ni, 35% Co and 1% Al. Two particle types, a flake shaped layered $LiCoO_2$ particle, and a spherical $LiNiO_2$ particle was identified on the cathode by EDX during the SEM analysis. Based on these two analyses, we conclude that the active cathode material is either an $NCA/LiCoO_2$ mixture, a so-called blend material, or a mixture of $LiNiO_2$ and $LiCoO_2$. As the Al content identified by ICP-OES can also be caused by contamination during the separation of the active cathode material and the aluminum current-collector, both material combinations are possible.

3.2. Post-Mortem Analysis

In this section, we first present the results of the optical PMA of the entire electrode surfaces. Figure 3 shows all electrode and separator sheets. Each number in Figure 3 corresponds to a galvanic element, 23 for the entire cell. Since the cell has been opened fully charged, most anodes are fully lithiated and, therefore, gold. As a first peculiarity in Figure 3, it is noticeable that in contrast to all other anodes, anode 1.8 and 1.12 is not fully lithiated. In the middle of the anode and cathode sheets, the pressure point of the sphere is visible. Abnormality of color can find in and around the pressure point at the anode, where a silver ring is visible. According to Cannarella et al. [4], this ring suggests the deposition of metallic lithium. This empirical observation matches the model of lithium deposition during cell charging of Tang et al. [8]. In the following, we will discuss all peculiarities of the consequence of inhomogeneous pressure on the cell stack.

Figure 4 shows three 1.8, 1.12, and 2.1 galvanic elements where the top cathode corresponds to the bottom anode. These galvanic elements are optically very different from the others in the cell after the experiment. Only the pressure point in the middle had been active Figure 4b,d,f, recognizable by the silver-colored ring in the middle of the anode as has been observed for all anodes. Also, the four dots, where the separator sticks together with the electrodes, in the corners are lithiated Figure 4b,d recognizable on the gold color around the dots. As a result of this, these dots stick on both electrodes and are therefore still active compared to the rest of the electrode. During the disassembly of the stack, no peculiarity regarding the contact of these anodes to the cathode was noticeable. A possible reason for the lack of lithiation could be that the pressure of the sphere caused the cell stack to bend so that the electrochemical contact of this electrode pair was lost during the cycling. Especially the points in Figure 4d, show a wave structure in the lithiation between the dots. Such wave structures can be created by bending and holding on to several points, so that the separator is folded in this region.

Figure 3. Optical post-mortem analysis after cell opening (**a**) stack 1 top (**b**) stack 2 bottom.

The galvanic element 1.12 in Figure 4c,d, which consists of an anode from stack one and a cathode from stack 2, also shows a very inhomogeneous lithium distribution around the pressure point. A possible reason for this is the transition from stack 1 to stack 2. The center of the anode 1.12 is lithiated more than the edge areas, which can be explained by the higher contact pressure in the center. In stack 2, all layers are lithiated similarly, and the pressure points have a similar appearance as well. The anodes of stack two elements 2.1 to 2.3 show an additional silver-colored diagonal line within the silver-colored ring in Figure 4f, which is very pronounced at 2.1 and almost entirely decreases up to 2.3. Also, the pressure point on the separator of galvanic element 2.1 is much darker than the other pressure point of the separator. For cathode layers 1.8 and 1.12 in Figure 4a,c, a silver-colored circle is also visible within the pressure area. The authors have no explanation for this coloration, and a change in color due to lithiation of the cathode is unknown in the literature, and the ICP-OES analysis shows no abnormality.

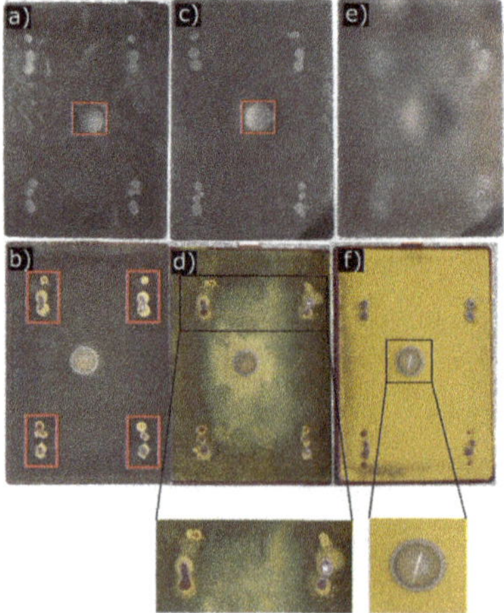

Figure 4. Optical PMA (**a,d**) galvanic element 1.8 (**b,e**) galvanic element 1.12 (**c,f**) galvanic element 2.1.

Figure 5 shows an example of an anode, cathode, and separator. The anode has a length of 60 mm and a width of 83.5 mm; compared to the cathode in Figure 5b, the anode is 1 mm bigger in all dimensions. Therefore, the edge of the anode becomes electrochemically inactive and does not participate in the charging and discharging process. For this reason, there is a black edge here called anode overhang. This anode overhang of the negative electrode beyond the edge of the additional positive capacitance and prevents lithium plating from occurring before the cutoff potential is reached [8].

The pressure point on the anode Figure 5a shows the anode with different color tones that corresponds to a different state of charges, and a silver-colored ring with the inner radius R_{inner} and the thickness of 1.8 mm, which indicates where the high ion flux starts and leads to lithium plating. This high current density rapidly builds a lithium metal layer on the particle surface. In this region, the anode voltage vs. Li/Li+ decreases to negative values, and metallic lithium deposited on the anode particles. Inside of this outer border region, a second ring with a very dark electrode surface has formed. As the dark color corresponds to a low state of charge, this area seems to be less active. Within the second ring, there is also a gold shimmer. One possible explanation is that the inner area was still active during cycling due to very inhomogeneous local pressure in this area. The non-transparent separator in this area supports this possibility. Nevertheless, the diffusion of lithium from active electrode parts outside the pressure area into the pressure area cannot be excluded. The initial cell voltage was 3.55 V, which corresponds to a SOC of 22%, where the graphite anode has not yet changed color [18]. Therefore, this area was lithiated during cell cycling.

The pressure point at the separator in Figure 5c has the same radius R_{out} as R_{inner}, which indicates the beginning of a transition zone of weaker pore closure, where a very high ion current density occurs. The intact separator outside the pressure area has a white color, and in the innermost edge of the pressure area, the separator becomes transparent. This transparent separator appears dark in the pictures presented here as the background behind the separator was black when the pictures made. The transparent separator indicates a pore closure as shown by Cannarella et al. [4]. Furthermore, in the pressure area, the separator becomes brighter again, where it assumed that the pores have not completely closed and that a part can still be active.

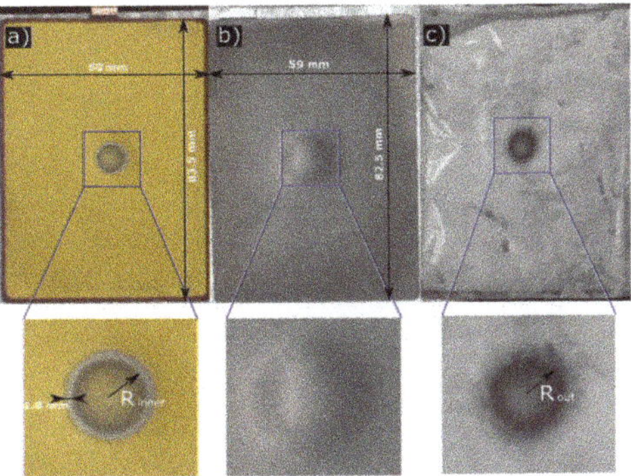

Figure 5. Exemplary layers of (**a**) Anode (**b**) Cathode (**c**) Separator.

3.3. Scanning Electron Microscope Analysis

To get a deeper understanding of the influence of pressure on active and passive materials on a microscopic level, the two cells were analyzed by SEM. The effects of homogeneously pressed dried electrodes compared to the effects of heterogeneously pressed cyclical cells should give an idea of how strong the effects could be. Since the pressure was adjusted so that the active materials are not damaged according to [15,16] and the experiments on the single electrode sheets, the particles should occur without damage. Considering the applied pressure is above 10 MPa, an irreversible pore closure should occur in the [2]. Before the SEM analysis, the separator was sputtered with silver atoms to prevent static charging during measurement. For the cathode and anode, this is not necessarily due to their electrical conductivity. Figure 6 shows an SEM image of the separator in the area below the sphere, direct at the boundary, and in an area without applied pressure area. In Figure 6b,c, the open pores of the separator are visible, here the separator is fully intact. Due to the uniform lithiation in this area, it is assumed that the ion current density had to be homogeneously distributed in this area. Figure 6d,e shows a region with closed pores and only a small region with open pores as marked red in Figure 6e. This mixed region of open and closed pores confirms the observation of the PMA that the region inside the silver ring with the gold shimmer was at a higher state of charge compared to the initial state. In the dark region at the boundary of the pressure area, the pores fully closed as shown in Figure 6f,g. This dark region of fully closed pores proves that the pressure was strong enough to close the pores, which results in a very uneven ion current density distribution so that locally very high ion current densities occur. This heterogeneous pore structure leads to an uneven current distribution of the electrochemical system. According to Tang et al., this is strongly influenced by the balance of ohmic resistance from the electrolyte and kinetic resistance at the electrode interface, which is influenced by diffusion in the graphite. Uneven current distributions influence the possibility of lithium plating because of the generation of overpotential [8]. This analysis coincides with the observation in the PMA, where lithium deposition was observed in the form of a silver ring in all anodes.

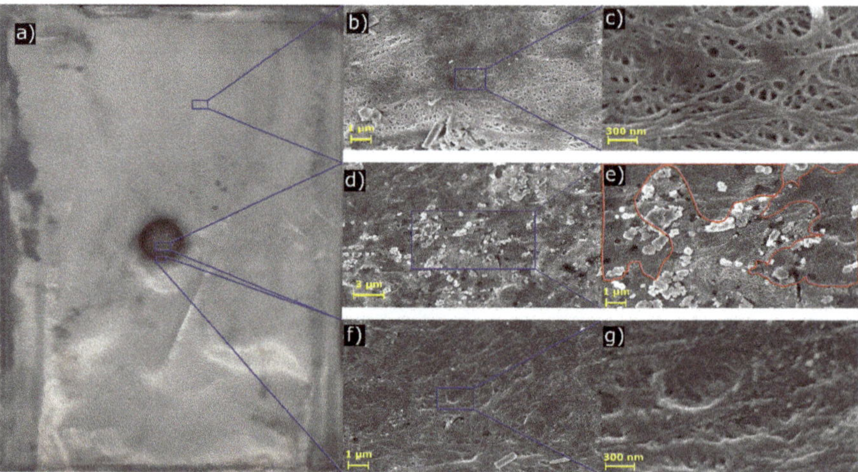

Figure 6. SEM image from separator (**a**) General view (**b**,**c**) unpressed region (**d**,**e**) pressed region in the center under the sphere (**f**,**g**) edge of the pressed region under the sphere.

Cannarella et al. builds a model of separator pore closure within a LIB based on Tang et al. [4,8]. The results of the model match with the observations of our experiment. The simulation from Cannarella et al. predicts a ring-shaped shape of the current density distribution around the pore-closed area on the negative electrode surface. Furthermore, by including the lithium plating kinetics and varying the corresponding plating exchange current density, the model calculates that the local potential of the anode vs. Li/Li+ will be negative around the pore-closed area so that the deposition of lithium will occur. This negative potential of the anode leads to lithium plating on the anode. Lithium plating is likely to lead to increased localized mechanical stress as this deposition can lead to measurable changes in anode thickness [20]. This local increase in thickness causes additional mechanical stress and could lead to further deformation of the components in the cell. Even if the simulation of Canarella et al. [4] is only a rough estimate of the real phenomena since the kinetics of lithium plating on graphite electrodes is still poorly understood. This work here and the simulation complement each other very well and jointly support the theory and understanding of the kinetics of lithium plating.

Previous publications of Cannarella et al., Lui et al. and Peabody et al. [2,4,5] mainly investigated the influence of defects in the separator, a question that remains to be clarified: What influence does the pressure have on the active materials in a commercial cell under cycling conditions? Figure 7 shows an SEM image of the cathode. The SEM image in the pressure region Figure 7b–d shows that the pressure was so high that the particles were strongly deformed that contradicts the pressure data of Tran et al. [16] and our experiments conducted on the uncycled single electrode sheets. In contrast, in these analyses, the active material is soaked with electrolyte and (de)lithiated during the application of pressure, which induces additional higher pressure on the particle due to the volume expansion of the graphite. Therefore, it assumed that both effects could have strongly influenced on the elastic properties of the particles. Figure 7c shows that the most NCA particles crushed and the $LiCoO_2$ flake in Figure 7d decompose in the direction of their layer structure. These cracks expose new surfaces to the electrolyte and enables an electrochemical degradation of the electrolyte, which introduced at the particle–electrolyte interface, the so-called cathode–electrolyte interface (CEI) [21]. The analogous process, as with graphite anodes, can cause lithium loss and gas formation.

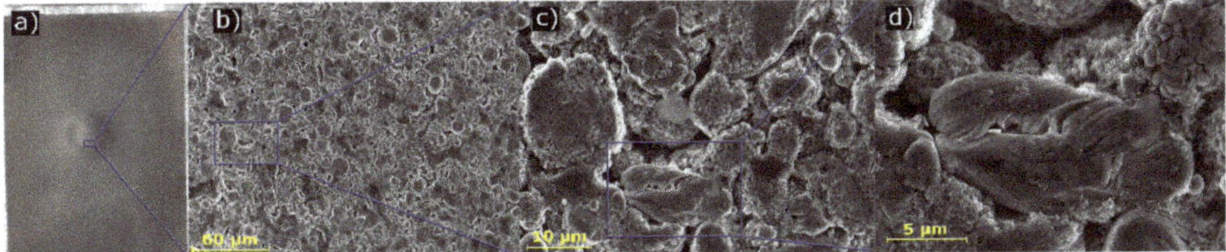

Figure 7. SEM image of cathode (**a**) general view of the cathode (**b–d**) pressed region.

In the SEM analysis of the anode, two regions in Figure 8a analyzed, as these differ significantly from their morphology and structure. In Figure 8b the border of the two regions is marked with a green line. The images of the anode in the pressure region in Figure 8f–h show that the contrast of the image deteriorates due to electrostatic charging of the sample during the SEM. It seems as if a new insulating layer has formed here, since the samples are electrostatically charged during the SEM. The influence on the electrochemical kinetic could come from squeezing the electrolyte partly away into the peripheral area. From this analysis, it cannot be determined whether the mechanical load deformed the graphite flakes. A possible cause for the insulating covering layer could be the growth of the solid–electrolyte interface (SEI). This layer can grow when new particle surfaces created by particle cracking exposed to the electrolyte. Microscopic analysis of the silver ring in Figure 8c–e shows a fine structure compared to the two previously discussed regions. This microstructure could be lithium dendrites or reaction products with deposited lithium. Unfortunately, metallic lithium cannot be determined with EDX and remains unexplained for the time being.

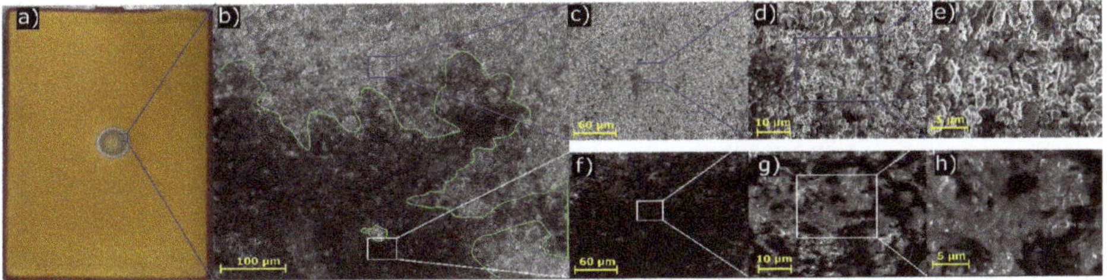

Figure 8. SEM image of the anode (**a**) overview (**b**) boundary of the silver ring to the pressure area (**c–e**) internal pressed region (**f–h**) transition zone (silver ring).

An uncycled cell was analyzed microscopically using SEM to compare pristine active materials with pressed active materials from the cycled cell. Furthermore, a dried single electrode layer of anode and cathode was put under a homogeneous pressure of 30 MPa, which is higher as the pressure in the experiment with the sphere, in order to compare dried pressed electrodes with soaked, cycled electrodes from the same LIB type. In Figure 9a–c, the particles of the pristine lithiated cathode are depicted. Figure 9d–f depicts the pressed lithiated cathode, which is like the pristine cathode. No deformed or crushed particles were observed. Also, the anode in Figure 10 shows no significant change under a homogeneous pressure of 30 MPa.

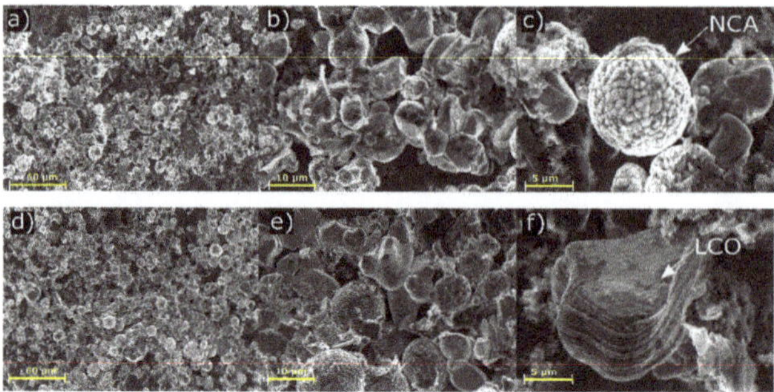

Figure 9. SEM image of a uncycled cell (**a–c**) overview of a pristine lithiated cathode (**d–f**) pressed area of a single lithiated cathode layer with 30 MPa.

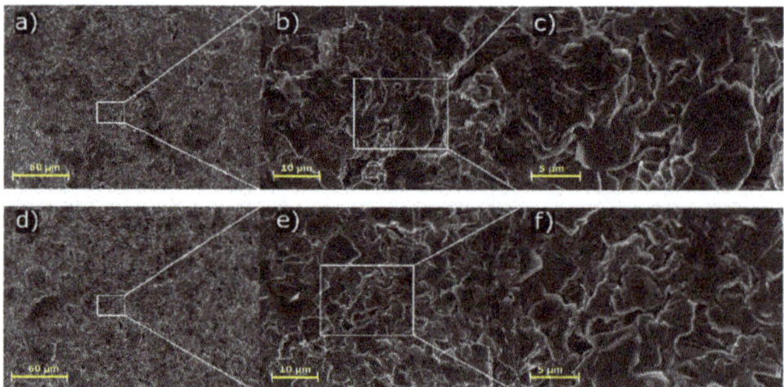

Figure 10. SEM image of a uncycled cell (**a–c**) overview of a pristine delithiated anode (**d–f**) pressed area of a single delithiated anode layer with 30 MPa.

4. Conclusions

In this work, a sphere was used as a force transmitter to induce an inhomogeneous pressure distribution on a cycled LIB. The sphere shape results in a heterogenic pressure distribution on the LIB, which has its maximum in the middle point and decreases towards the edge and induce a ring of locally high electrochemical activity that leads to lithium plating. The force of 1.4 kN, which results in a maximum pressure of 20 MPa, was chosen so that only the separator closes its pores, and no influence on the positive and negative electrode should take place. The experiment in this thesis has shown that pressures in this low range are sufficient to cause defects in the active materials.

A macroscopic PMA of the cell shows that local mechanical effects change the properties of different components. Very inhomogeneous state of charge on individual layers show that these lose contact from each other in the multi-layer system. This contact loss indicates an internal bending of the layers due to different mechanical stress. Moreover, a discoloration was found on the individual two cathodes for which there is no explanation. Furthermore, the PMA showed that the load on the separator at the pressure point was very different, which could be seen optically by different color changes or transparencies. The edge of the pressure point had the most substantial color change, which indicates a secure pore closure of the separator. The center indicated a still partially active region. A silver ring of metallic lithium around the pressure point found on all anodes during the PMA. The lithium plating ring had a width of 1.8 mm and began where the substantial color change of the separator ended. A high current density due to the pore closure of the separator generating an overpotential in this region and create conditions which favor plating created the lithium plating ring.

A microscopic examination at selected locations on the separator, anode, and cathode confirmed and showed new findings regarding the effect of the pressure point on the LIB. The examination of the separator confirmed that below the pressure area, the separator was still partially active, as partly closed and open pores found in this region. The edge area of the separator showed closed entirely pores, which led to the uneven current distribution. The findings of the cathode show that almost all particles were crushed and deformed under pressure. Although the chosen maximum pressure should theoretically be too low for this, therefore, there must have been additional pressure development. The additional pressure development during (de)lithiation of the particles can be an explanation for the extreme increase in pressure. Since the graphite expands by 5% to 11% [17,22] and the cathode expands by 1% to 2% [17,23] the increase in pressure is probably caused by the anode over the cycling.

Moreover, it assumed that the fact that the particles were soaked with electrolyte has changed their elastic properties. Therefore, the authors assume that the pressure influence of dry active materials cannot be transferred to battery cells with filled electrolytes and has a strong influence. The findings of the anode show a transition of the pressure point to the lithium plating ring. In the area of the pressure area, a new insulating layer has formed on the particles because the SEM images have become blurred due to electrostatic charging. In the area of the lithium plating ring, the morphology has changed. This microstructure could be lithium dendrites or reaction products with deposited lithium.

Overall, homogeneous pressure distribution on LIB must be considered during the battery module design, especially pressure points with small areas should be prevented that can lead to unexpected phenomena. This pressure points can result in high-pressure development within the LIB during operation, causing lithium plating and dendrite growth, which can cause short circuits by penetrating the thin separator and trigger a thermal runaway.

Author Contributions: Conceptualization, G.F. and D.U.S.; Formal analysis, G.F. and F.R.; Methodology, G.F.; Validation, G.F.; Investigation, G.F. and L.W.; Writing—original draft, G.F.; Writing—review & editing, G.F., L.W., F.R. and D.U.S.

Funding: This research received no external funding.

Acknowledgments: The authors would like to thank Philipp Wunderlich (Institute of Inorganic Chemistry) for the SEM images EDX analysis of the probes and Rita Graff for the ICP-OES analysis. Philipp Dechent for proofreading.

Conflicts of Interest: The authors declare no conflict of interest.

Abbreviations

The following abbreviations are used in this manuscript:

LIB	Lithium-Ion Battery
EDP	Electrolyte Degradation Product
LEI	Lithium–Electrolyte Interface
PMA	Post-Mortem Analysis
SEM	Scanning Electron Microscope
EDX	Energy Dispersive X-ray Spectroscopy
ICP-OES	Inductively Coupled Plasma-Optical Emission Spectrometry

EC	Ethylene Carbonate
EMC	Ethylmethyl Carbonate
LiPF6	Lithiumhexafluorophosphat
LiCoO2	Lithium-Cobalt(III)-oxid
LiNiO2	Lithium Nickel Oxide
NCA	Lithium Nickel Cobalt Aluminum Oxide (LiNiCoAlO$_2$)
NMC	Lithium Nickel Cobalt Manganese Oxide (LiNiCoMnO$_2$)
SOC	State of Charge

References

1. Cannarella, J.; Arnold, C.B. State of health and charge measurements in lithium-ion batteries using mechanical stress. *J. Power Sources* **2014**, *269*, 7–14. [CrossRef]
2. Peabody, C.; Arnold, C.B. The role of mechanically induced separator creep in lithium-ion battery capacity fade. *J. Power Sources* **2011**, *196*, 8147–8153. [CrossRef]
3. Cannarella, J.; Arnold, C.B. Ion transport restriction in mechanically strained separator membranes. *J. Power Sources* **2013**, *226*, 149–155. [CrossRef]
4. Cannarella, J.; Arnold, C.B. The Effects of Defects on Localized Plating in Lithium-Ion Batteries. *J. Electrochem. Soc.* **2015**, *162*, A1365–A1373. [CrossRef]
5. Liu, X.M.; Fang, A.; Haataja, M.P.; Arnold, C.B. Size Dependence of Transport Non-Uniformities on Localized Plating in Lithium-Ion Batteries. *J. Electrochem. Soc.* **2018**, *165*, A1147–A1155. [CrossRef]
6. Oh, K.Y.; Siegel, J.B.; Secondo, L.; Kim, S.U.; Samad, N.A.; Qin, J.; Anderson, D.; Garikipati, K.; Knobloch, A.; Epureanu, B.I.; et al. Rate dependence of swelling in lithium-ion cells. *J. Power Sources* **2014**, *267*, 197–202. [CrossRef]
7. Wünsch, M.; Kaufman, J.; Sauer, D.U. Investigation of the influence of different bracing of automotive pouch cells on cyclic liefetime and impedance spectra. *J. Energy Storage* **2019**, *21*, 149–155. [CrossRef]
8. Tang, M.; Albertus, P.; Newman, J. Two-Dimensional Modeling of Lithium Deposition during Cell Charging. *J. Electrochem. Soc.* **2009**, *156*, A390. [CrossRef]
9. Rahe, C.; Kelly, S.T.; Rad, M.N.; Sauer, D.U.; Mayer, J.; Figgemeier, E. Nanoscale X-ray imaging of ageing in automotive lithium ion battery cells. *J. Power Sources* **2019**, *433*, 126631. [CrossRef]
10. Waldmann, T.; Gorse, S.; Samtleben, T.; Schneider, G.; Knoblauch, V.; Wohlfahrt-Mehrens, M. A Mechanical Aging Mechanism in Lithium-Ion Batteries. *J. Electrochem. Soc.* **2014**, *161*, A1742–A1747. [CrossRef]
11. Bach, T.C.; Schuster, S.F.; Fleder, E.; Müller, J.; Brand, M.J.; Lorrmann, H.; Jossen, A.; Sextl, G. Nonlinear aging of cylindrical lithium-ion cells linked to heterogeneous compression. *J. Energy Storage* **2016**, *5*, 212–223. [CrossRef]
12. Waldmann, T.; Hogg, B.I.; Wohlfahrt-Mehrens, M. Li plating as unwanted side reaction in commercial Li-ion cells—A review. *J. Power Sources* **2018**, *384*, 107–124. [CrossRef]
13. Su, X.; Dogan, F.; Ilavsky, J.; Maroni, V.A.; Gosztola, D.J.; Lu, W. Mechanisms for Lithium Nucleation and Dendrite Growth in Selected Carbon Allotropes. *Chem. Mater.* **2017**, *29*, 6205–6213. [CrossRef]
14. Loveridge, M.; Remy, G.; Kourra, N.; Genieser, R.; Barai, A.; Lain, M.; Guo, Y.; Amor-Segan, M.; Williams, M.; Amietszajew, T.; Ellis, M.; Bhagat, R.; Greenwood, D. Looking Deeper into the Galaxy (Note 7). *Batteries* **2018**, *4*, 3. [CrossRef]
15. Wang, C.W.; Yi, Y.B.; Sastry, A.M.; Shim, J.; Striebel, K.A. Particle Compression and Conductivity in Li-Ion Anodes with Graphite Additives. *J. Electrochem. Soc.* **2004**, *151*, A1489. [CrossRef]
16. Tran, H.Y.; Greco, G.; Täubert, C.; Wohlfahrt-Mehrens, M.; Haselrieder, W.; Kwade, A. Influence of electrode preparation on the electrochemical performance of LiNi0.8Co0.15Al0.05O2 composite electrodes for lithium-ion batteries. *J. Power Sources* **2012**, *210*, 276–285. [CrossRef]
17. Rieger, B.; Schlueter, S.; Erhard, S.V.; Schmalz, J.; Reinhart, G.; Jossen, A. Multi-scale investigation of thickness changes in a commercial pouch type lithium-ion battery. *J. Energy Storage* **2016**, *6*, 213–221. [CrossRef]
18. Shellikeri, A.; Watson, V.; Adams, D.; Kalu, E.E.; Read, J.A.; Jow, T.R.; Zheng, J.S.; Zheng, J.P. Investigation of Pre-lithiation in Graphite and Hard-Carbon Anodes Using Different Lithium Source Structures. *J. Electrochem. Soc.* **2017**, *164*, A3914–A3924. [CrossRef]

19. Uhlmann, C.; Illig, J.; Ender, M.; Schuster, R.; Ivers-Tiffée, E. In situ detection of lithium metal plating on graphite in experimental cells. *J. Power Sources* **2015**, *279*, 428–438. [CrossRef]
20. Bitzer, B.; Gruhle, A. A new method for detecting lithium plating by measuring the cell thickness. *J. Power Sources* **2014**, *262*, 297–302. [CrossRef]
21. Liu, Y.M.; G Nicolau, B.; Esbenshade, J.L.; Gewirth, A.A. Characterization of the Cathode Electrolyte Interface in Lithium Ion Batteries by Desorption Electrospray Ionization Mass Spectrometry. *Anal. Chem.* **2016**, *88*, 7171–7177. [CrossRef] [PubMed]
22. Hahn, M.; Buqa, H.; Ruch, P.W.; Goers, D.; Spahr, M.E.; Ufheil, J.; Novák, P.; Kötz, R. A Dilatometric Study of Lithium Intercalation into Powder-Type Graphite Electrodes. *Electrochem. Solid-State Lett.* **2008**, *11*, A151. [CrossRef]
23. Itou, Y.; Ukyo, Y. Performance of LiNiCoO$_2$ materials for advanced lithium-ion batteries. *J. Power Sources* **2005**, *146*, 39–44. [CrossRef]

 *sustainability*

Article

High-Precision Monitoring of Volume Change of Commercial Lithium-Ion Batteries by Using Strain Gauges

Lisa K. Willenberg [1,2], Philipp Dechent [1,2], Georg Fuchs [1,2], Dirk Uwe Sauer [1,2,3,*] and Egbert Figgemeier [2,4,*]

1   Institute for Power Electronics and Electrical Drives (ISEA), RWTH Aachen University, 52066 Aachen, Germany; lisa.willenberg@isea.rwth-aachen.de (L.K.W.); Philipp.Dechent@isea.rwth-aachen.de (P.D.); Georg.Fuchs@isea.rwth-aachen.de (G.F.)
2   Juelich Aachen Research Alliance, JARA-Energy, 52062 Aachen, Germany
3   Institute for Power Generation and Storage Systems (PGS), RWTH Aachen University, 52062 Aachen, Germany
4   Forschungszentrum Juelich GmbH, IEK-12, Helmholtz-Institut Münster, co ISEA of RWTH Aachen University, 52066 Aachen, Germany
*   Correspondence: batteries@isea.rwth-aachen.de (D.U.S.); egbert.figgemeier@isea.rwth-aachen.de (E.F.)

Received: 17 December 2019; Accepted: 9 January 2020; Published: 11 January 2020

Abstract: This paper proposes a testing method that allows the monitoring of the development of volume expansion of lithium-ion batteries. The overall goal is to demonstrate the impact of the volume expansion on battery ageing. The following findings are achieved: First, the characteristic curve shape of the diameter change depended on the state-of-charge and the load direction of the battery. The characteristic curve shape consisted of three areas. Second, the characteristic curve shape of the diameter change changed over ageing. Whereas the state-of-charge dependent geometric alterations were of a reversible nature. An irreversible effect over the lifetime of the cell was observed. Third, an s-shaped course of the diameter change indicated two different ageing effects that led to the diameter change variation. Both reversible and irreversible expansion increased with ageing. Fourth, a direct correlation between the diameter change and the capacity loss of this particular lithium-ion battery was observed. Fifth, computer tomography (CT) measurements showed deformation of the jelly roll and post-mortem analysis showed the formation of a covering layer and the increase in the thickness of the anode. Sixth, reproducibility and temperature stability of the strain gauges were shown. Overall, this paper provides the basis for a stable and reproducible method for volume expansion analysis applied and established by the investigation of a state-of-the-art lithium-ion battery cell. This enables the study of volume expansion and its impact on capacity and cell death.

Keywords: lithium-ion batteries; ageing; mechanical fatigue; volume change; measurement technique; silicon anode

---

1. Introduction

Mechanical characteristics of lithium-ion battery cells are of major importance when designing applications with maximized energy density and lifetime [1,2]. With the inherent volume variations during the use of lithium-ion batteries, the integration of cells into modules, packs, and systems needs to account for geometric variations and the induced mechanical stresses changing with state-of-charge and state-of-health [3,4]. In this context, an inappropriate design will lead to premature ageing of components on all levels. In extreme cases, a mismatch of volume requirements and device integration can lead to catastrophic failure [5].

In the past, chemical and electrochemical ageing effects of lithium-ion batteries have been the major focus in the international research community. Topics such as side reactions, the solid electrolyte interphase (SEI) growth, loss of lithium inventory, and separator clogging have been investigated in great detail [6–9]. The influence of mechanical ageing has often been ignored and is now attracting attention with the aim of the highest energy densities and the introduction of alloying materials with high volume variations during charging and discharging [3]. Moreover, complex system integration in electric vehicles requires detailed understanding and quantitative prediction of geometric variations as a function of state-of-charge and state-of-health.

Mechanical ageing might be one of the reasons for early and unexpected cell death, especially considering that graphite can expand up to 7–12% [3,10,11], Lithium Nickel Cobalt Aluminium Oxide (NCA) up to 5% [3], Nickel Manganese Cobalt (NMC) 1–2% [3,12,13], and silicon up to 280% [14] within the given voltage limits. Thus, the question to be clarified is how serious is the influence of mechanics on cell ageing in commercially relevant cells with respect to its used case?

The volume change can have various causes: The first is lithium migration, in which electrode materials change in volume as a result of lithium intercalation and deintercalation into their crystal structures [14–18]. In addition, the gas formation can occur due to side reactions [6,19,20]. Furthermore, it is known that thick layers of electrolyte decomposition products are formed as an almost impermeable covering layer [6,21,22]. Alternatively, lithium plating can also occur, resulting in swelling behavior [6,23,24].

It has already been shown by other measurement methods, such as thickness gauge [3,25–27], pressure [3], digital image correlation, [28] and multi-scale investigation [11], that a volume change takes place, which is strongly material dependent and measures between 3 and 10% for a full cell [3,11,25–28] between charge and discharge.

For our study, we investigated an ageing matrix consisting of 51 Samsung 35E battery cells. Twelve different ageing conditions were applied and a minimum of three batteries per ageing point were tested. At three ageing points, the C-rate was additionally varied with two batteries each. Before the data of all batteries can be evaluated in detail, the measuring method must be verified in a first step. Because long time frames of the study, in combination with expected effects in the lower micrometer range, it desires a careful evaluation of the method. In this context, the current study evaluates the application of strain gauges as a valuable tool to monitor geometric changes in cylindrical cells with high precision. Firstly, the validity of the strain gauge is evaluated by investigating the signal drifts as a function of temperature and time. Secondly, the results from one battery are presented as an illustrative example. Finally, a post-mortem investigation was performed. The data of the strain gauge, in combination with electrochemical characterization, gives information regarding the lithium-ion battery and their degradation as well as parameters for battery system designs.

2. Materials and Methods

2.1. Investigated Lithium-Ion Battery

A commercially available lithium-ion battery, Samsung SDI INR18650 35E, was selected as the test object to check the functionality of the strain gauge. Verification measurements and ageing measurements were carried out on this lithium-ion battery.

The battery's specifications are given in Table 1. The investigated battery was a high-energy lithium-ion battery with a usable voltage range of 2.65–4.20 V according to the datasheet of the producer [29].

Table 1. Specification of the investigated lithium-ion battery [29].

	Samsung SDI INR18650 35E
Diameter	Max. Φ 18.55 mm
Length w/o terminals	Max. 65.25 mm
Cell weight	50.00 g max
Standard discharge capacity	Min. 3350.00 mAh
Charging voltage	4.20 V
Discharge cut-off voltage	2.65 V
Max. charge current	2.00 A
Max. discharge current	8.00 A (for continuous discharge) 13.00 A (not for continuous discharge)
Operating temperature (cell surface temperature)	Charge: 0 to 45 °C Discharge: −10 to 60 °C
Initial internal impedance	Initial internal impedance measured at AC 1kHz after Standard charge. Initial internal impedance $\leq$ 35 mΩ

Computer tomography (CT) measurements were performed in order to obtain more information about the geometry and construction details of the investigated battery cell. A Werth TomoScope HV Compact (Werth Messtechnik GmbH, Gießen, Germany) was used with a microfocus transmission tube with up to 225 kV. All shown images have a resolution of 38 μm/voxel. The visualization software myVGL 3.2.5 (Volume Graphics GmbH, Heidelberg, Germany) was used for analyzing the data.

One Samsung 35E lithium-ion battery was opened under an argon atmosphere. A remaining voltage of 2.60 V was measured directly before opening. Spatially distributed discs of the double-coated electrodes, with a diameter of 20 mm, were taken. As preparation, each sample was washed with Dimethylcarbonat (Dimethylcarbonat Msynth®plus, Merck KGaA, Darmstadt, Germany) before it was dissolved in aqua regia. The solution was filled with distilled water until a 100 ml solution was obtained. Induced coupled plasma-optical emission spectrometer (ICP-OES) was conducted on this solution. The electrode compositions of anode and cathode were measured with Varian 725 induced coupled plasma-optical emission spectrometer (Agilent, Santa Clare, USA). The electrodes and the separator were evaluated by their surface morphology and color using a Keyence VK-9710 laser microscope (KEYENCE DEUTSCHLAND GmbH, Neu-Isenburg, Germany).

Electrical Tests

For examination and ageing of the lithium-ion battery, the cell test system Digatron MCFT 20-05-50ME with the precision of the current measurement of 0.2% (Digatron Power Electronics GmbH, Aachen, Germany) was used. The temperature sensor Dallas Semiconductor DS18B20 (Maxim Integrated, San Jose, USA) with a precision of ±1 °C was located on the lithium-ion battery case close to the strain gauge.

The lithium-ion battery test was conducted in a temperature chamber Binder MK 720 (BINDER GmbH, Tuttlingen, Germany) with a temperature precision of ±2 °C. The experiment was conducted at a constant ambient temperature of 25 °C. However, the lithium-ion battery temperature may have changed due to self-heating depending on the current.

The battery was electrically aged by charge-discharge cycling. The ageing cycles were performed at 25 °C and a mean state of charge (SOC) of 50%. A cycle depth of 50% was chosen. The lithium-ion battery was charged with 0.5 C to the upper charging voltage limit and discharged Ah-based to avoid voltage drift.

In order to determine the ageing of the lithium-ion battery, check-ups were performed at 25 °C. Check-ups were done every 50th equivalent full cycle and included capacity determination, pulse resistance, and quasi-open-circuit voltage (quasi-OCV) characteristics. During the check-up, the lithium-ion battery was first discharged down to 2.65 V, followed by a pause of 10 min. Then the battery was charged with 0.5 C (1.7 A) up to 4.20 V, followed by constant-voltage charging up to

I < 0.02C (maximum 180 min). Afterwards, the lithium-ion battery was kept at rest for 30 min before the capacity was determined with a discharge of 0.5C until the voltage of 2.65 V was reached, followed by a pause of 10 min. Next, the lithium-ion battery was charged, followed by a pause of 30 min.

The quasi-OCV was measured with a current of 0.1C. For this, the lithium-ion battery was discharged down to 2.65 V and charged up to 4.20 V with a current of 0.1C, followed by a pause of 30 min.

2.2. Strain Gauges

The strain gauge 1-LY11-6/120A from HBM Germany (Hottinger Baldwin Messtechnik GmbH, Darmstadt, Germany) was attached to the lithium-ion battery housing with superglue Z70 (Hottinger Baldwin Messtechnik GmbH, Darmstadt, Germany) according to the enclosed manufacturer's instructions. It was assumed that the height (z-direction) of the battery does not change, as there is enough buffer in the housing in this direction. Therefore, it was assumed that only the diameter will change. In order to measure the change in diameter without the influence of edge effects, the strain gauge was placed in the middle of the housing height in the circumferential direction.

The area and the measuring grid area of the strain gauge were 6×13 mm^2 and 6×2.7 mm^2, respectively. The resistance of the strain gauge was the electrical resistance between the two connecting cables. The change in volume of the measuring body caused a strain of the measuring grid. Due to the strain of the measuring grid, the resistance of the measuring grid changed [30]. This resistance change induced a signal which was amplified by a measuring amplifier Q.bloxx A116 120/350 and converted into a strain ε in µm/m by Gantner Q.station 101DT (Gantner Instruments Test & Measurement GmbH, Darmstadt, Germany) according to Equation (1).

$$\varepsilon = \frac{signal}{k} \times \frac{4}{F} \times 1000, \tag{1}$$

The signal is the measured signal of the strain gauge. The k factor is the characteristic value of a strain gauge. The value here is $k = 2.04$. The bridge factor F indicates how many active strain gauges are present in the Wheatstone bridge. For a quarter bridge, the manufacturer indicated a value of one.

The resulting diameter change of the cylindrical lithium-ion battery is calculated as follows: It is assumed that the elongation of the lithium-ion battery in the circumferential direction is constant over the height of the casing. The change in diameter can be inferred from the elongation of the circumference using Equation (2).

$$\Delta d = \frac{U_0 + U_0 \times (\varepsilon - \varepsilon(1))}{\pi} - d_0, \tag{2}$$

Δd is the diameter change of the lithium-ion battery. U_0 in mm represents the circumference at the start of the test. U_0 is calculated using Equation (3). ε is the measured strain in µm/m and d_0 is the diameter at the start of the test. In our case, this corresponds to 18.5 mm.

$$U_0 = \pi \times d_0, \tag{3}$$

3. Results

3.1. Analysis of Cylindrical Cell

The Samsung 35E was analyzed by ICP-OES. The anode consists of graphite and a small fraction of silicon was detected. The cathode consists out of nickel-cobalt-aluminum with a high amount of nickel as the cathodic active material.

The construction of the Samsung 35E lithium-ion battery cell consists of a jelly roll with a mandrel in the centre, see Figure 1. The electrode is wound 19 times around the mandrel. The current tab of the anode is located at the outer edge of the jelly roll, see Figure 1a. The cathode current tab is placed in the middle of the jelly roll. The diameter of the entire lithium-ion battery is given as a maximum value of 18.55 mm, according to the datasheet [29]. For the calculation using Equation (2), 18.5 mm was used.

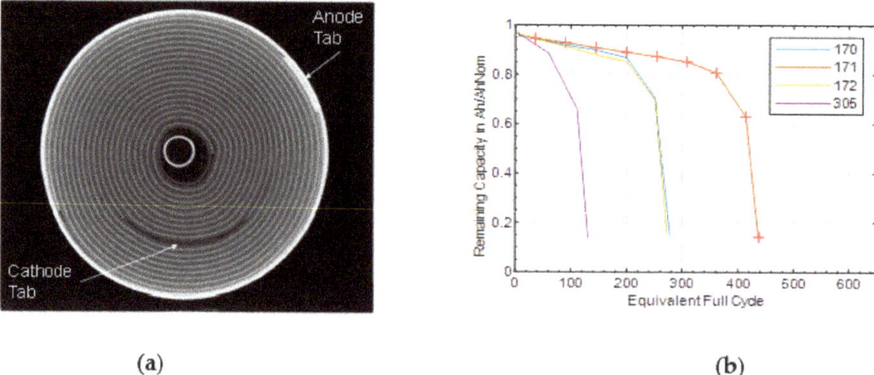

(a)
(b)

Figure 1. (**a**) Computer tomography (CT) measurement of an as-received Samsung 35E (347) with the location of the electrode tabs. (**b**) Remaining capacity over equivalent full-cycle measured during check-ups with a current of 0.5 C of Samsung 35E (170, 171, 172, 305) of the electrical ageing at a mean state of charge (SOC) of 50% and a cycle depth of 50%. Each cell was displayed in a different color. The red crosses mark the time of the check-ups.

Figure 1b shows the remaining capacity of four lithium-ion batteries over equivalent full cycles. The cycle ageing was performed at a mean SOC of 50% and a cycle depth of 50% for all four battery cells. As can be seen, these lithium-ion battery cells aged linearly first at a rate of 0.5C, which was above the recommended maximum C-rate in order to accelerate ageing effects. Afterwards, the remaining capacity dropped strongly. All four cells showed a fast cell death. The reason for the fast cell death may be the increased C-rate of 0.5C. The manufacturer recommends 1/3C. If this battery is aged with 1/3C, it reaches clearly higher cycle numbers of more than 1000 equivalent full cycles. Furthermore, rather large variations in ageing characteristics can be seen at this ageing point. This fits the rather large variations in ageing characteristics, which had been observed for this battery type in the entire ageing matrix. Since these results are not the focus of this paper, they will be described in another paper. In the following, only the battery Samsung 35E 171 will be discussed to focus on the topic of this publication.

3.2. Validation of the Measurement Process

The strain ε was measured according to Equation (1) and is indicated in µm. According to the manufacturer, the maximum extensibility of the sensor is 50,000 µm/m [30]. The maximum strain change in our measurement was 1000 µm/m per full cycle of a lithium-ion battery. The exact maximum diameter change depends on the preload of the strain gauge. With an alternating strain of $\varepsilon_W = \pm 1000$ µm/m, a zero-point change of $\varepsilon_m \Delta \leq 30$ µm/m can be expected with more than 10^7 load cycles [30]. With our measurements, we perform 100 load cycles per 50 equivalent full cycles (cycle depth of 50%). This resulted in 100,000 equivalent full cycles until we exceed the limit of 10^7 load cycles confirming the suitability of the strain gauge for the current task.

To investigate the volume change of the battery, its diameter change was calculated using Equation (2). Figure 2a shows the diameter change drift depending on the temperature at five different temperatures for more than 20 h. The strain gauge was attached as described in chapter 2.2. As can be seen, the resulting diameter change was stable over the relevant time period.

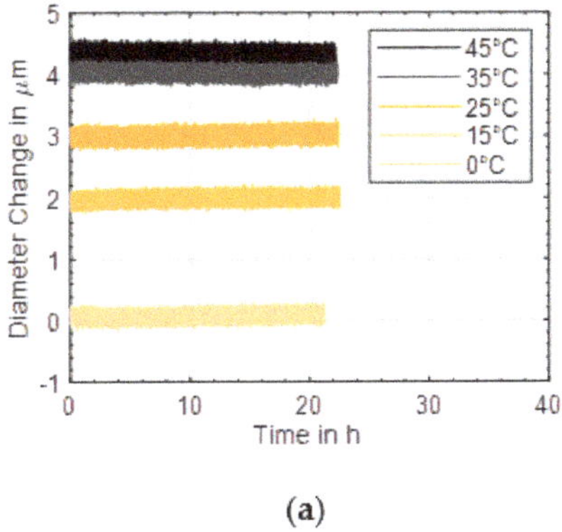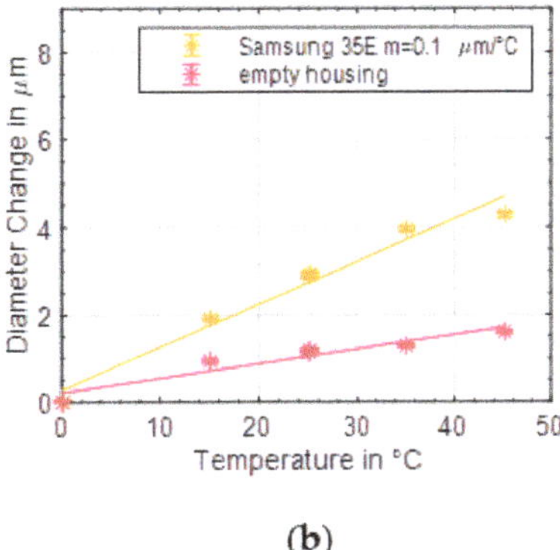

(a)　　　　　　　　　　　　　　(b)

Figure 2. (**a**) Diameter change of a lithium-ion battery cell (197) at different temperatures. (**b**) Derivation of the temperature dependence of the Samsung 35E (197) and of an empty battery housing with a strain gauge including a standard deviation over the measured 20 h.

Based on these results (Figure 2a), a relationship between diameter change of the lithium-ion battery together with the strain gauge and the temperature can be derived, see Figure 2b. The strain gauge and the Samsung 35E together are changing their diameter by 0.1 μm/°C.

In order to determine the temperature influence of the jelly roll (including electrode stack and electrolyte), the diameter change of a battery housing without jelly roll was measured, see Figure 2b. The strain gauge and the empty battery housing together are changing their diameter below the limits of detection.

In order to investigate the temperature cycle stability of the strain gauge, a Samsung 35E lithium-ion battery and an empty housing with strain gauges were temperature cycled between 0 and 40 °C. Figure 3 shows an extract of the measurement. In total, 15 temperature cycles were performed. Every ten hours, the temperature was changed in the climate chamber. Figure 3 shows two temperature cycles of 15 as an example.

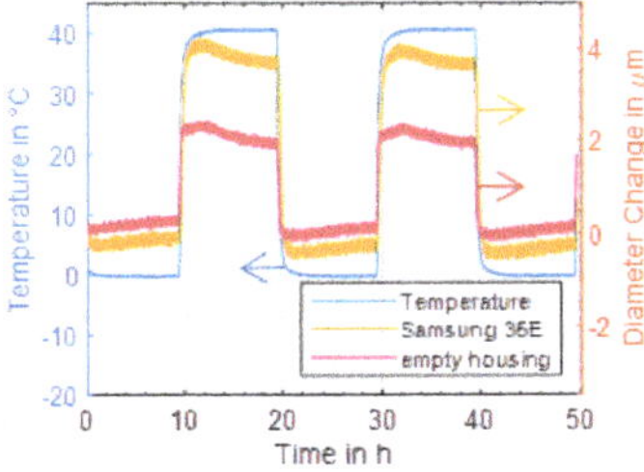

Figure 3. Measurement of the diameter change on a Samsung 35E (197) lithium-ion battery during temperature cycling between 0 and 40 °C. The temperature of the climate chamber was changed every 10 h.

The temperature cycle stability is shown in Figure 4 based on the measurement shown in Figure 3. In Figure 4a the mean results and the standard deviation of 40 °C of the Samsung 35E and the empty housing are shown. The mean results and standard deviations were calculated from the results shown in Figure 3. At 40 °C, the mean diameter change of the empty housing was stable with a maximum standard deviation of 0.1 μm. The mean diameter change of the Samsung 35E was decreasing with the temperature cycle number with a maximum standard deviation of 0.33 μm. Figure 4b shows the mean results and the standard deviation at 0 °C. At 0 °C, the same behavior can be observed. The maximum deviation of the Samsung 35E and the empty housing was 0.1 μm and 0.1 μm, respectively.

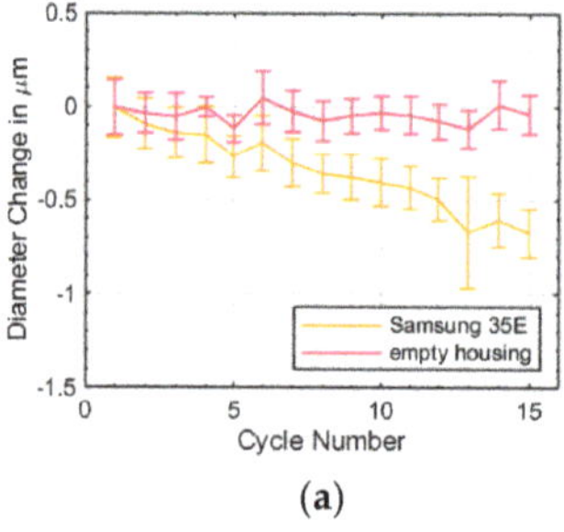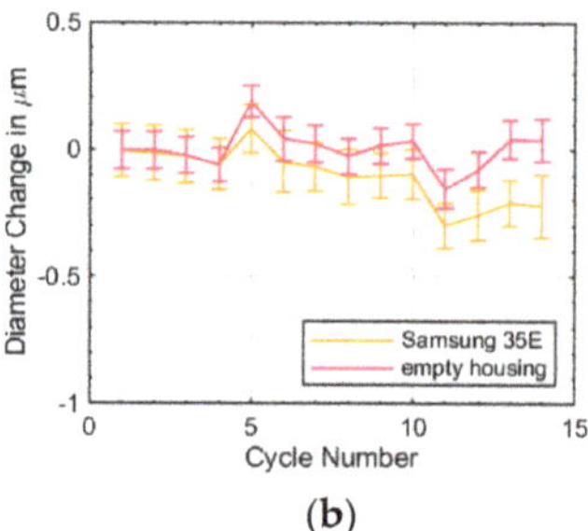

(a)   (b)

Figure 4. Measurement of the diameter change on a lithium-ion battery during temperature cycling between 0 and 40 °C. The temperature of the climate chamber was changed every 10 h. (**a**) shows the mean temperature and mean diameter change with standard deviation at 40 °C of the lithium-ion battery Samsung 35E and an empty housing (average of the numbers between 35 and 40 °C) and (**b**) at 0 °C (average of the numbers between 0 and 5 °C).

3.3. Diameter Change of Samsung 35E During Electrical Ageing

Figure 5 shows the discharge voltage profile and the associated change in diameter over the state of discharge for a lithium-ion battery cell as-received.

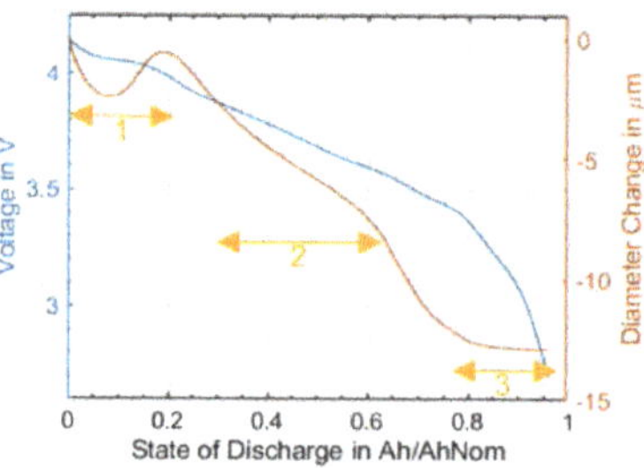

Figure 5. Diameter change and voltage profile from a 0.1C discharge between 4.2 and 2.65 V of an as-received lithium-ion battery at 25 °C (171).

At the beginning of the discharge, the diameter change decreases until it reached a minimum of at approximately 10% state of discharge. Then, the volume rised again up to a maximum at 20% state of discharge in area 1 in Figure 5. The maximum was below the diameter at the beginning of the measurement. Afterwards, the diameter change decreased again. Between a state of discharge of 40%–60%, the diameter change was flatter than before. An area with a constant gradient can be seen in area 2. From 60% state of discharge onwards, the diameter change decreased again until it flattened out, starting from 75% onwards and remained constant until the lithium-ion battery was discharged in area 3.

Figure 5 shows a representative measurement. Overall, 51 specimens were tested. The experiments showed a diameter change of around 10.7 µm with a standard deviation of 4.4 µm for an as-received lithium-ion battery during a full discharge process.

For the comparison of charge and discharge, the discharged lithium-ion battery was normalized to SOC of 0%. In a check-up, the battery was first discharged to 0% SOC and then charged to 100% SOC.

When comparing the diameter change of a charge and discharge of a lithium-ion battery, the shape differed, see Figure 6a. When the lithium-ion battery was charged, its change in diameter increases directly from 0% onwards, see Figure 6b. When it is discharged, the range up to 10% SOC was rather flat. The area with a constant gradient between 40% and 60% SOC was more pronounced with the charge than in the discharge, see area 2. This is where the deviation was most noticeable. During discharge, the slope at 30% SOC was greater than during charge. Furthermore, the turning point became less evident at 60% SOC. The charge and discharge curve cross each other at 30% and 70% SOC. Afterwards, both curves take the same course. The minimum at high SOCs is showing the same for both the charge and the discharge, see area 1.

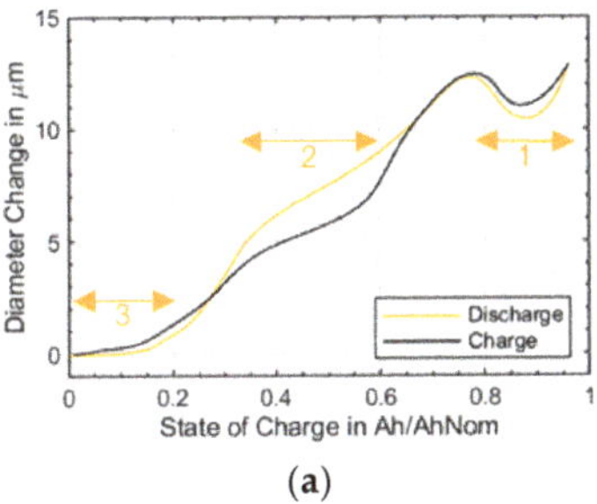
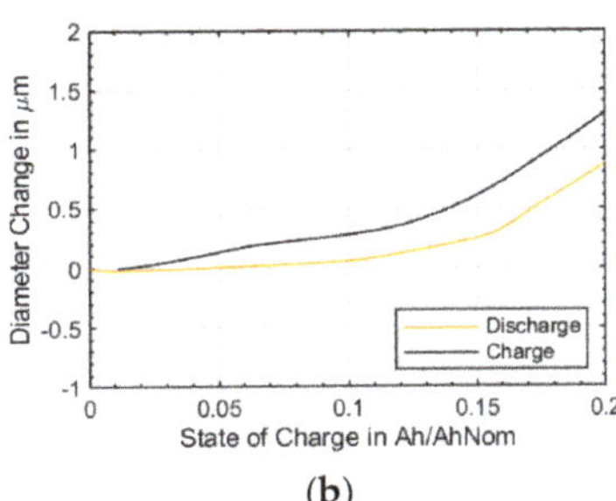

Figure 6. Diameter change of the Samsung 35E (171) during discharge and charge between 4.2 and 2.65 V at 0.1C and 25 °C (**a**) between stage of charge 0–1 Ah/Ah$_{Nom}$ and (**b**) enlargement between stage of charge 0–0.2 Ah/Ah$_{Nom}$.

Figure 7 shows the diameter change due to ageing as a function of SOC and capacity (a) discharge and (b) charge. Each curve represents one check-up performed on the same lithium-ion battery. All curves were normalized to the diameter of the beginning of life check-up.

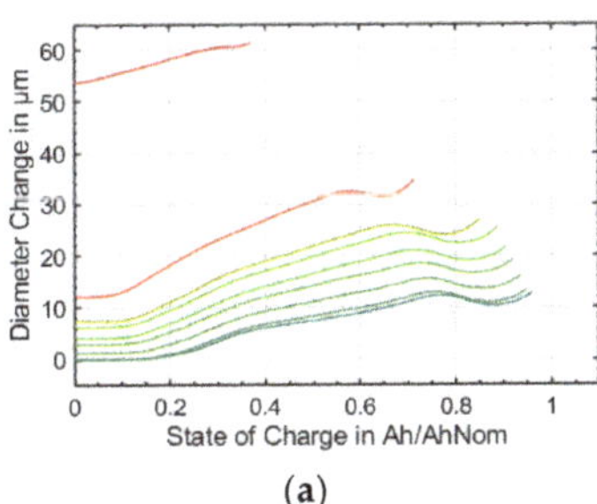
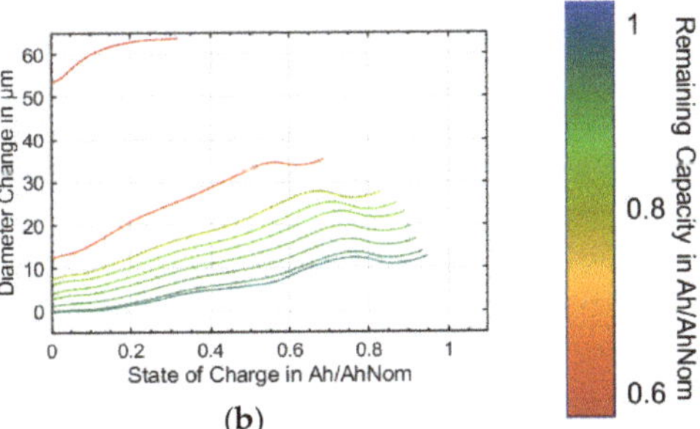

Figure 7. Diameter change measured with 0.1C over SOC at the check-ups for (**a**) discharge and (**b**) charge. The lithium-ion battery (171) was aged at a mean SOC of 50% and a cycle depth of 50% with 0.5C.

The shape of the curves remains nearly the same until a remaining capacity of 80%. Then the diameter change shrinks and the areas 1 and 3 start to disappear. In addition, the distance between the curves increases. After the seventh check-up the lithium-ion battery suddenly suffered a rapid capacity loss, see Figure 1b. This was accompanied by a significant increase in diameter. In addition, the area with a constant gradient in area 2 is no longer clearly visible. There were no major differences between charging and discharging during ageing. Only the last check-up showed a clearly different curve. During discharge, the change in diameter increased linearly and the minimum was still recognizable at the end of the curve. During charge, the diameter change followed an s-shape curve.

Figure 8 shows the diameter change over ageing during discharge at "SOC 0%", "SOC 100%", and the "difference" between both. Two different volume change mechanism can be observed: First, a reversible diameter change caused by charging and discharging (lithiation and delithiation of the electrodes), represents here as the "difference" between diameter at "SOC 0%" and diameter at "SOC 100%". The reversible diameter change "difference" increased with ageing until the end of life of the lithium-ion battery. Here, the diameter change difference was decreasing again. Secondly, an irreversible diameter change can be observed. It was clearly distinguishable with progressing ageing: The diameter change values at "SOC 0%" and "SOC 100%" change and induced a moving "difference" during ageing which is reversible. The irreversible diameter change, the basic level of the lithium-ion battery diameter rised with ageing, both "SOC 0%" and "SOC 100%". At the end of life, the lithium-ion battery had a diameter change of above 60 μm in the fully charged state (SOC 100%).

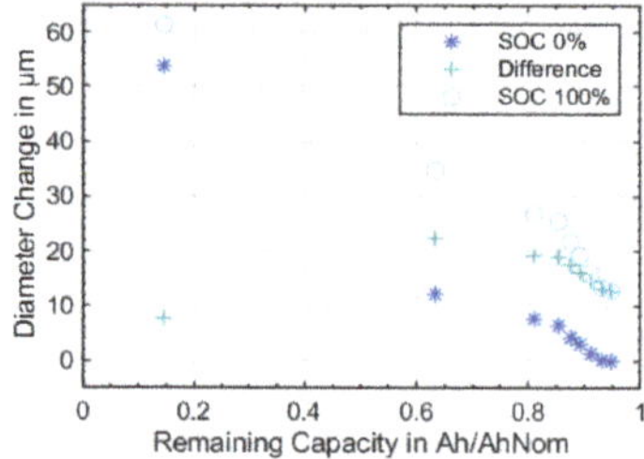

Figure 8. Diameter change measured with 0.1C over remaining capacity at SOC 0%, SOC 100%, and the difference for discharge. The lithium-ion battery (171) was aged at a mean SOC of 50% and a cycle depth of 50% with 0.5C.

In addition, it was noticeable that the diameter change curves took an s-shaped course up to a remaining capacity of 80%, see Figure 8. Thereafter, the expansion increased linearly. The results of the charge behaved the same as the discharge.

Looking at the increase of the mean diameter change not during the check-ups but during the cyclisation, it can be seen that the mean diameter change increased slowly up to a cycle number of about 700 cycles, shown in Figure 9. Afterwards there was a steep increase in the mean diameter change. This process fits very well to the previously shown diameter changes during the check-ups, shown in Figures 7 and 8. There are small increases in the mean diameter change due to the check-ups.

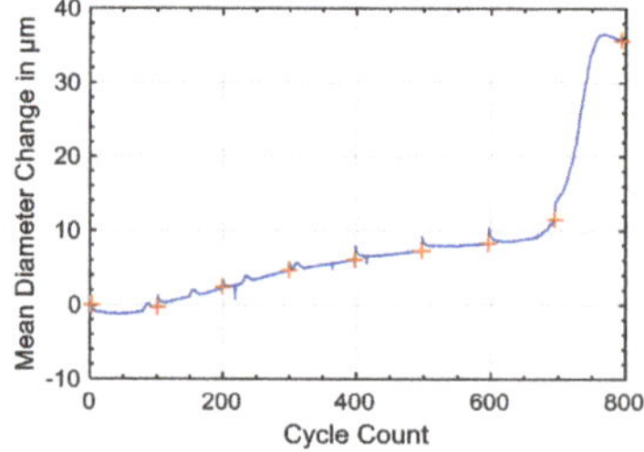

Figure 9. Mean diameter change of the Samsung 35E (171). For the diameter change curve, the mean diameter change from the ageing cycles (SOC of 50% and a cycle depth of 50% with 0.5C) during charging was used. Therefore, this was applied over cycles and not equivalent full-cycles. The red crosses mark the time of the check-ups.

The temperature changes during quasi-OCV was less than 1 °C. During an entire check-up, the temperature changed from 25 °C to a maximum of 30 °C during the capacity test. As can be seen inFigure 10, the outside mean temperature of the lithium-ion battery during each cycle rised constantly until a cycle count of 600. Afterwards, the mean temperature increased faster until the end of life. However, 30 °C was not exceeded during the entire aging process.

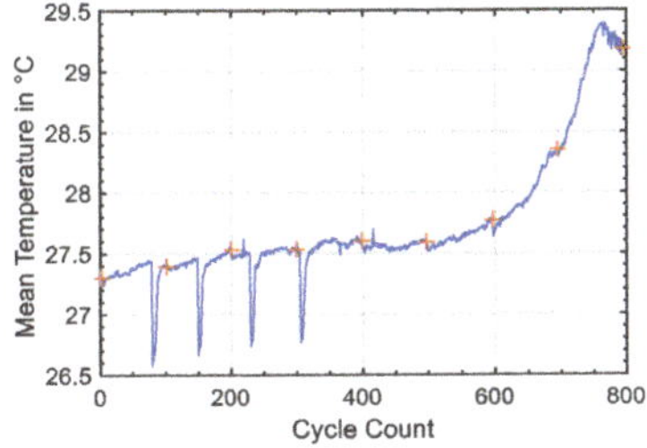

Figure 10. Mean temperature of the Samsung 35E (171). For the temperature curve, the mean temperature from the ageing cycles (SOC of 50% and a cycle depth of 50% with 0.5C) during charging was used. Therefore, this was applied over cycles and not equivalent full-cycles. The spikes interrupting the trend are caused by an opening of the temperature chamber or by other batteries aged in the same climate chamber. The red crosses mark the time of the check-ups.

3.4. Post-Mortem Study of Samsung 35E

Figure 11b shows a CT measurement of the aged Samsung 35E (171) at a remaining capacity of 0.1 Ah/Ah$_{Nom}$. The deformation of the jelly roll is clearly visible. The deformation occurs only in the middle of the jelly roll and especially in the middle of the cell height (z-direction). For the middle five windings, deformation can be observed.

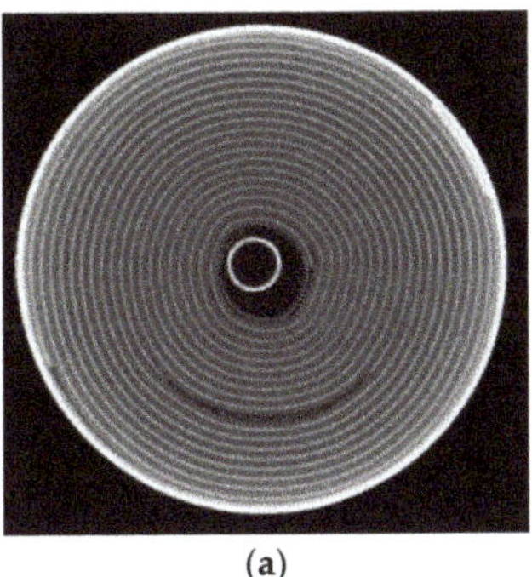

(a)

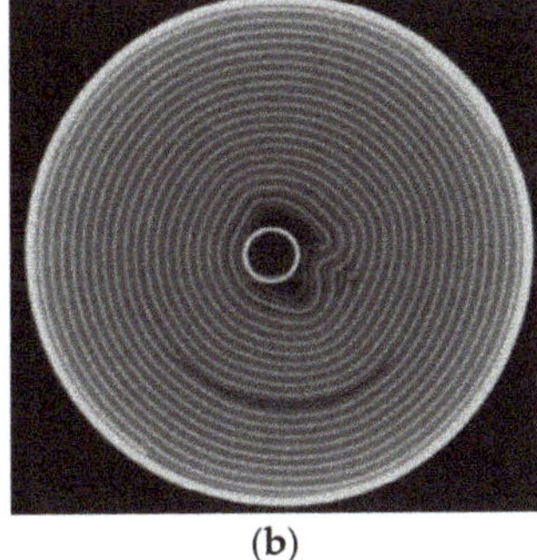

(b)

Figure 11. CT measurement of (**a**) an as-received Samsung 35E (347) and (**b**) the aged Samsung 35E (171) at a remaining capacity of 0.1 Ah/Ah$_{Nom}$. Winding deformations in the middle of the jelly roll are clearly visible.

When the aged battery was opened, clear deformations were visible in the cathode and anode, as can be seen in Figure 12a,b. In addition, a grey covering layer was formed, especially in the middle part of the jelly roll. This part started after the cathode tab. The position of the cathode tab was clearly visible on the anode due to the complete absence of the coating and the visible copper current collector (Figure 12c).

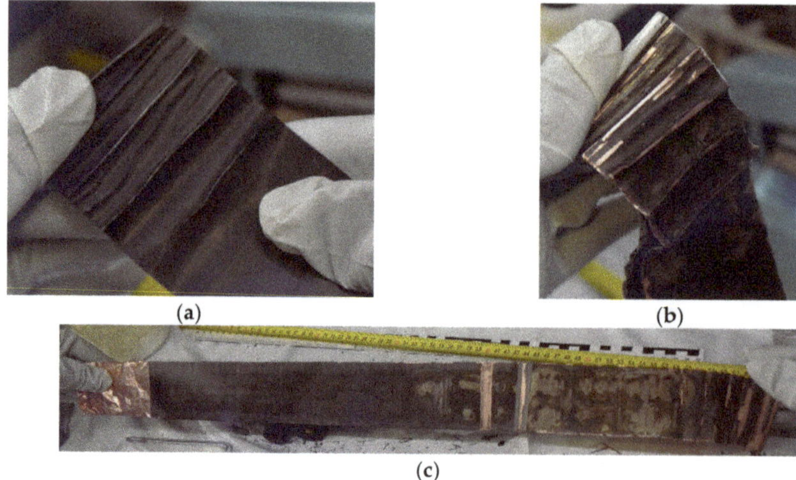

Figure 12. Pictures of electrode coatings of an aged Samsung 35E (171), (a) cathode and (b,c) anode after cell opening. Deformation and delamination are visible in each winding in the middle of the jelly roll. On the anode, a grey colored covering layer is visible.

By analyzing the electrodes with a laser microscope, it can be shown that a covering layer was formed on the anode and cathode. On the anode, the coating was not quite as pronounced in the black area (Figure 13b) as it was in the grey area (Figure 13c). However, only a few particles were still visible as in the as-received state. The particles were still visible at the cathode. Only a light covering layer was formed here (Figure 13e).

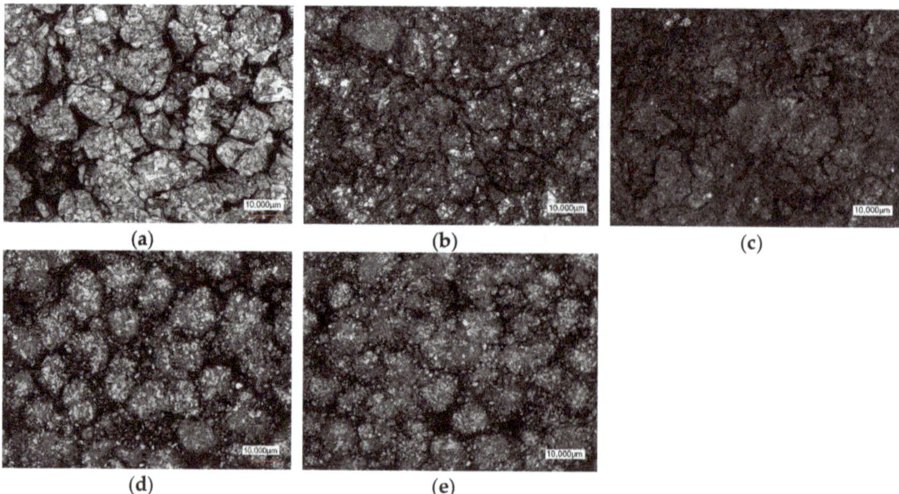

Figure 13. Laser microscope picture of the Samsung 35E, (a) anode as-received (347), (b) aged anode of a black spot of Figure 12c and (c) aged anode of a grey spot of Figure 12c (171), (d) cathode as-received (347) and (e) aged (171).

4. Discussion

As could be shown in the results section, a strain gauge is a very stable and reliable measuring tool. During cycling of the Samsung 35E, the maximum permissible limit of the sensor, 50,000 µm/m [30], was not exceeded until the end-of-life of the battery. Herein, a drift of ≤30 µm/m can be assumed at a cycle number of more than 100,000 equivalent full cycles. The lithium-ion battery considered here fell below the end of life capacity of 60% as given by the producer [29] already after 420 equivalent full cycles. Therefore, it can be assumed that the drift of the strain gauge was low enough for a precise time series measurement. This finding was confirmed with the investigations of the temperature influence, see Figure 2. Here, the diameter change of the Samsung 35E was measured at different temperatures for more than 20 h and the temperature influence was determined. It turned out that the measurement yields stable results at all temperatures. In addition, a linear temperature influence of the battery and strain gauge of 0.1 µm/°C could be demonstrated, which renders the temperature impact on measured results neglectable. Furthermore, it could be shown, that the expansion of the jelly roll due to temperature change had only a minor impact. Thus, the influence of temperature can be neglected. Since the lithium-ion battery did not heat up to more than 30 °C during ageing, the temperature influence can be limited to a maximum of 0.5 µm (Figure 10). This can be considered as minor compared to the measured diameter changes of more than 10 µm during charge and discharge.

As shown in Figure 4 the mean diameter change of the Samsung 35E was decreasing with temperature cycle number at 0 and 40 °C. The reason for this decreasing might be a reduction of the mechanical stress within the jelly roll or a homogenization of the electrodes. This behavior requires further investigation. The decrease in the diameter change showed that the irreversible diameter change from Figure 7 is caused by effect within the battery and that the diameter of the battery was increasing over ageing.

The diameter change of the discharge process was around 10.7 µm with a standard deviation of 4.4 µm for an as-received lithium-ion battery. Rather large variations in battery production and ageing characteristics could be the reason for the large deviation in diameter change [31]. Further results illuminated this aspect and will be reported elsewhere.

If one considers the diameter change during discharging (Figure 5), a characteristic trend can be seen: It can be divided into three areas. A minimum at high SOC, an area with a constant gradient in medium SOC region, and a flattening slope towards the end of discharge. The impact of the electrode material of this effect is still under investigation, but relating our results with the literature suggests that the effect at low SOC might be caused by the anode [3]. Furthermore, there was little volume change in the region of medium SOC because of the stage transition of graphite (stage transition: 2L→2 [32]), which shows a plateau type of characteristic when relating to the state of charge and volume [3].

Due to the differences in diameter change that have been observed between charge and discharge, it can be assumed that intercalation and de-intercalation induce asymmetric volume changes at anode and cathode (Figure 6).

As can be seen in Figure 7, the shape of the curves remained nearly the same until a remaining capacity of 80%. Then, the diameter changes receded and the characteristic shape of the curve at the beginning of charge and end of charge started to disappear. In addition, the area with a constant gradient in medium SOC range was no longer clearly visible. This might be explained by a loss of active material or active lithium which can cause less active volume overall that is used for charging and discharging or due to the fact that the electrodes have shifted to each other so that not all areas of the electrodes are used anymore. Pore clogging may lead to an increase in the reversible diameter change of the battery. In this case, the pores would no longer be available as void space for accommodating the increase in the volume of the particle during intercalation, so the diameter change of the active material was translating directly into diameter change of the cylindrical lithium-ion battery as monitored by the strain gauges.

Figure 4 shows that the irreversible diameter change stems from the battery and not from the strain gauge characteristic. Hence, the reasons for the irreversible diameter change could be an increase of the SEI, the non-uniform formation of a covering layer or lithium plating, or gassing [19–24]. The post-mortem results showed that a covering layer was formed over aging (Figures 12 and 13). The exact separation between SEI growths, plating or covering layer formation is unfortunately not possible. In addition, a thickening of the anode could also be observed, which fits in with the formation of the covering layer. The thickness of the anode changed from 174 μm (347) to 211 μm (171). The thickness of the cathode stayed constant with 156 μm.

The s-shaped course of the diameter change described in Figure 8 indicated two different effects that led to the diameter change variation. The s-shape course was completed from a remaining capacity of 80%. Afterwards, the change in diameter increased linearly. Which effects played a role here is to be clarified further. The s-shaped course existed up to a remaining capacity of 80%, which correlated well with the already shown roll over the capacity (Figure 1b).

Furthermore, the irreversible diameter change accelerated significantly at lower states-of-health (Figures 7 and 8). After check-up seven at around 420 equivalent full cycles, the lithium-ion battery suddenly lost more capacity from cycle to cycle. This was accompanied by a sudden increase in the diameter, see Figure 9. It can be assumed that due to the expansion of the anode, the pressure in the housing was increased to such an extent that the winding was deformed to reduce the pressure (Figure 11). However, the pressure increased both during the charging and discharging cycles as well as during ageing due to the ageing effects described.

At this point, it is not possible to say whether the jelly roll deformation is the reason for the sudden cell death. However, the deformation of the jelly roll could be a reason for the increasing diameter change.

Overall, there was a direct correlation between the diameter change of the lithium-ion battery and the capacity. The precise interplay between mechanistic, intercalation driven, and chemical effects deserves further investigation. The precise interplay between mechanistic, intercalation driven, and chemical effects deserves further investigation. Nevertheless, Dahn et al. [33] have shown that macroscopic deposition layers were an indicator of capacity roll-over, which was in agreement with our results.

The results can be summed up as follows: It is important to consider the volume expansion when designing and ageing lithium-ion batteries. Therefore, the strain gauges is a stable and reproducible method to analyze the volume expansion and its impact on capacity and cell death.

5. Conclusions

Electrochemical failure mechanisms like SEI growth, formation of a covering layer, plating, or gas formation induce mechanical stress and therefore a volume expansion. All factors increasing the probability for sudden death of the battery cell. A straightforward and reliable method for measuring the volume expansion and their influence on the state-of-health of the lithium-ion battery is the strain gauge. As we demonstrated, it yielded stable and reproducible results, when monitoring geometric variations on the micrometer scale.

The diameter change of the lithium-ion battery described a characteristic trend and is SOC and load direction-dependent. For as-received batteries of type Samsung 35E, a mean diameter change of 10.7 μm during a full discharge process was measured. Different phenomena seemed to play a role in the ageing of the battery so that the change in diameter changed as a function of the ageing of the battery. Two effects can be observed over ageing using strain gauges: First, an increase of irreversible diameter change, and second, an increase of reversible diameter change. Both reversible and irreversible expansion increased with ageing. Overall, two different ageing effects led to the diameter change variation. CT measurements showed deformation of the jelly roll and post-mortem analysis showed the formation of a covering layer and the increase in the thickness of the anode.

Finally, direct correlation between the diameter change and the capacity loss of this lithium-ion battery can be identified. Therefore, the strain gauge is a tool for predicting the sudden cell death and might even serve as a diagnostic tool for the state-health, including safety aspects.

Author Contributions: Conceptualization, L.K.W., P.D., D.U.S., and E.F.; methodology, L.K.W., G.F., and P.D.; validation, L.K.W.; formal analysis, L.K.W.; investigation, L.K.W.; resources, D.U.S. and E.F.; data curation, P.D. and L.K.W.; writing—original draft preparation, L.K.W. and E.F.; writing—review and editing, L.K.W., P.D., G.F., E.F., and D.U.S.; visualization, L.K.W. and P.D.; supervision, D.U.S. and E.F.; funding acquisition, D.U.S and E.F. All authors have read and agreed to the published version of the manuscript.

Funding: The results were generated in the research training group "mobilEM" (GRK 1856/1, 1856/2). We would like to thank the Deutsche Forschungsgemeinschaft (DFG) for funding.

Acknowledgments: This work has been done during the time as an associated member of the research training group 'mobilEM' (GRK 1856/1, 1856/2). The authors would like to thank Marcel Eckert for the setup of the test bench; Rita Graff, Martin Graff, and Niklas Kürten for the cell opening and ICP-OES measurements; and Moritz Teuber for proofreading.

Conflicts of Interest: The authors declare no conflicts of interest.

References

1. Pfrang, A.; Kersys, A.; Kriston, A.; Sauer, D.U.; Rahe, C.; Käbitz, S.; Figgemeier, E. Long-term cycling induced jelly roll deformation in commercial 18650 cells. *J. Power Sources* **2018**, *392*, 168–175. [CrossRef]

2. Waldmann, T.; Gorse, S.; Samtleben, T.; Schneider, G.; Knoblauch, V.; Wohlfahrt-Mehrens, M. A Mechanical Aging Mechanism in Lithium-Ion Batteries. *J. Electrochem. Soc.* **2014**, *161*, A1742–A1747. [CrossRef]

3. Louli, A.J.; Li, J.; Trussler, S.; Fell, C.R.; Dahn, J.R. Volume, Pressure and Thickness Evolution of Li-Ion Pouch Cells with Silicon-Composite Negative Electrodes. *J. Electrochem. Soc.* **2017**, *164*, 2689–2696. [CrossRef]

4. Cannarella, J.; Arnold, C.B. Stress evolution and capacity fade in constrained lithium-ion pouch cells. *J. Power Sources* **2014**, *245*, 745–751. [CrossRef]

5. Loveridge, M.J.; Remy, G.; Kourra, N.; Genieser, R.; Barai, A.; Lain, M.J.; Guo, Y.; Amor-Segan, M.; Williams, M.A.; Amietszajew, T.; et al. Looking deeper into the galaxy (Note 7). *Batteries* **2018**, *4*, 3. [CrossRef]

6. Vetter, J.; Novák, P.; Wagner, M.R.; Veit, C.; Möller, K.C.; Besenhard, J.O.; Winter, M.; Wohlfahrt-Mehrens, M.; Vogler, C.; Hammouche, A. Ageing mechanisms in lithium-ion batteries. *J. Power Sources* **2005**, *147*, 269–281. [CrossRef]

7. Broussely, M.; Biensan, P.; Bonhomme, F.; Blanchard, P.; Herreyre, S.; Nechev, K.; Staniewicz, R.J. Main aging mechanisms in Li ion batteries. *J. Power Sources* **2005**, *146*, 90–96. [CrossRef]

8. Peabody, C.; Arnold, C.B. The role of mechanically induced separator creep in lithium-ion battery capacity fade. *J. Power Sources* **2011**, *196*, 8147–8153. [CrossRef]

9. Wohlfahrt-Mehrens, M.; Vogler, C.; Garche, J. Aging mechanisms of lithium cathode materials. *J. Power Sources* **2004**, *127*, 58–64. [CrossRef]

10. Zhang, N.; Tang, H. Dissecting anode swelling in commercial lithium-ion batteries. *J. Power Sources* **2012**, *218*, 52–55. [CrossRef]

11. Rieger, B.; Schlueter, S.; Erhard, S.V.; Schmalz, J.; Reinhart, G.; Jossen, A. Multi-scale investigation of thickness changes in a commercial pouch type lithium-ion battery. *J. Energy Storage* **2016**, *6*, 213–221. [CrossRef]

12. Bitzer, B.; Gruhle, A. A new method for detecting lithium plating by measuring the cell thickness. *J. Power Sources* **2014**, *262*, 297–302. [CrossRef]

13. Dolotko, O.; Senyshyn, A.; Mühlbauer, M.J.; Nikolowski, K.; Ehrenberg, H. Understanding structural changes in NMC Li-ion cells by in situ neutron diffraction. *J. Power Sources* **2014**, *255*, 197–203. [CrossRef]

14. Liu, X.H.; Wang, J.W.; Huang, S.; Fan, F.; Huang, X.; Liu, Y.; Krylyuk, S.; Yoo, J.; Dayeh, S.A.; Davydov, A.V.; et al. In situ atomic-scale imaging of electrochemical lithiation in silicon. *Nat. Nanotechnol.* **2012**, *7*, 749–756. [CrossRef] [PubMed]

15. Allart, D.; Montaru, M.; Gualous, H. Model of lithium intercalation into graphite by potentiometric analysis with equilibrium and entropy change curves of graphite electrode. *J. Electrochem. Soc.* **2018**, *165*, A380–A387. [CrossRef]

16. Dahn, J.R. Phase diagram of LixC6. *Phys. Rev. B* **1991**, *44*, 9170–9177. [CrossRef]

17. Kubota, Y.; Escao, M.C.S.; Nakanishi, H.; Kasai, H. Crystal and electronic structure of Li15Si4. *J. Appl. Phys.* **2007**, *102*, 053704. [CrossRef]

18. Li, J.; Dahn, J.R. An in situ X-ray diffraction study of the reaction of Li with crystalline Si. *J. Electrochem. Soc.* **2007**, *154*, 156–161. [CrossRef]

Sustainability **2020**, *12*, 557

19. Self, J.; Aiken, C.P.; Petibon, R.; Dahn, J.R. Survey of gas expansion in Li-Ion NMC pouch cells. *J. Electrochem. Soc.* **2015**, *162*, A796–A802. [CrossRef]

20. Kumai, K.; Miyashiro, H.; Kobayashi, Y.; Takei, K.; Ishikawa, R. Gas generation mechanism due to electrolyte decomposition in commercial lithium-ion cell. *J. Power Sources* **1999**, *81–82*, 715–719. [CrossRef]

21. Peled, E.; Menkin, S. Review-SEI: Past, present and future. *J. Electrochem. Soc.* **2017**, *164*, A1703–A1719. [CrossRef]

22. Lewerenz, M.; Warnecke, A.; Sauer, D.U. Post-mortem analysis on LiFePO4|Graphite cells describing the evolution & composition of covering layer on anode and their impact on cell performance. *J. Power Sources* **2017**, *369*, 122–132.

23. Petzl, M.; Kasper, M.; Danzer, M.A. Lithium plating in a commercial lithium-ion battery—A low-temperature aging study. *J. Power Sources* **2015**, *275*, 799–807. [CrossRef]

24. Legrand, N.; Knosp, B.; Desprez, P.; Lapicque, F.; Raël, S. Physical characterization of the charging process of a Li-ion battery and prediction of Li plating by electrochemical modelling. *J. Power Sources* **2014**, *245*, 208–216. [CrossRef]

25. Lee, J.H.; Lee, H.M.; Ahn, S. Battery dimensional changes occurring during charge/discharge cycles—Thin rectangular lithium ion and polymer cells. *J. Power Sources* **2003**, *119–121*, 833–837. [CrossRef]

26. Grimsmann, F.; Brauchle, F.; Gerbert, T.; Gruhle, A.; Knipper, M.; Parisi, J. Hysteresis and current dependence of the thickness change of lithium-ion cells with graphite anode. *J. Energy Storage* **2017**, *12*, 132–137. [CrossRef]

27. Oh, K.Y.; Siegel, J.B.; Secondo, L.; Kim, S.U.; Samad, N.A.; Qin, J.; Anderson, D.; Garikipati, K.; Knobloch, A.; Epureanu, B.I.; et al. Rate dependence of swelling in lithium-ion cells. *J. Power Sources* **2014**, *267*, 197–202. [CrossRef]

28. Luo, J.; Dai, C.Y.; Wang, Z.; Liu, K.; Mao, W.G.; Fang, D.N.; Chen, X. In-situ measurements of mechanical and volume change of LiCoO2 lithium-ion batteries during repeated charge–discharge cycling by using digital image correlation. *Measurement* **2016**, *94*, 759–770. [CrossRef]

29. Specification of Product for Lithium-Ion Rechargeable Cell Model Name: INR18650-25R. Available online: https://www.orbtronic.com/content/samsung-35e-datasheet-inr18650-35e.pdf (accessed on 22 October 2019).

30. Dehnungsmessstreifen Vollendete Präzision von HBM. Available online: https://www.me-systeme.de/produkte/dehnungsmessstreifen/catalog/s1264-DMS-Katalog-fuer-Spannungsanalyse.pdf (accessed on 22 October 2019).

31. Baumhöfer, T.; Brühl, M.; Rothgang, S.; Sauer, D.U. Production caused variation in capacity aging trend and correlation to initial cell performance. *J. Power Sources* **2014**, *247*, 332–338. [CrossRef]

32. Whittingham, M.S. Materials Challenges Facing Electrical Energy Storage. *MRS Bull.* **2008**, *33*, 411–419. [CrossRef]

33. Burns, J.C.; Kassam, A.; Sinha, N.N.; Downie, L.E.; Solnickova, L.; Way, B.M.; Dahn, J.R. Predicting and Extending the Lifetime of Li-Ion Batteries. *J. Electrochem. Soc.* **2013**, *160*, A1451–A1456. [CrossRef]

Article

A Comparative Study on the Influence of DC/DC-Converter Induced High Frequency Current Ripple on Lithium-Ion Batteries

Pablo Korth Pereira Ferraz *,† **and Julia Kowal** †

Electrical Energy Storage Technology, Department of Energy and Automation Technology, Technische Universität Berlin, Einsteinufer 11, D-10587 Berlin, Germany; julia.kowal@tu-berlin.de
* Correspondence: pablo.korthpereiraferraz@tu-berlin.de
† Current address: Electrical Energy Storage Technology, Departement of Energy and Automation Technology, Faculty IV, Secr. EMH 2, Technische Universität Berlin, Einsteinufer 11, D-10587 Berlin, Germany.

Received: 17 September 2019; Accepted: 28 October 2019; Published: 31 October 2019

Abstract: Modern battery energy systems are key enablers of the conversion of our energy and mobility sector towards renewability. Most of the time, their batteries are connected to power electronics that induce high frequency current ripple on the batteries that could lead to reinforced battery ageing. This study investigates the influence of high frequency current ripple on the ageing of commercially available, cylindrical 18,650 lithium-ion batteries in comparison to identical batteries that are aged with a conventional battery test system. The respective ageing tests that have been carried out to obtain numerous parameters such as the capacity loss, the gradient of voltage curves and impedance spectra are explained and evaluated to pinpoint how current ripple possibly affects battery ageing. Finally, the results suggest that there is little to no further influence of current ripple that is severe enough to stand out against ageing effects due to the underlying accelerated cyclic ageing.

Keywords: lithium-ion battery; battery ageing; cyclic ageing tests; current ripple; triangular current; power electronics; DC/DC-converter; half bridge converter; distribution of relaxation times

---

1. Introduction

Battery energy systems for stationary storages as well as for electric vehicles that are based on lithium-ion batteries are widely regarded as one of the key technologies to realize the energy transition [1]. A determined shift could mitigate the most severe aspects of climate change so that in recent years the research on battery materials, modelling, ageing, systems and integration has been thriving [2]. Yet, amongst others, a lot of open questions remain in the field of diagnosing batteries connected to power electronics in practical conversion systems although they are already comprehensively used in stationary systems and electric vehicles alike. Batteries, connected to power electronics that play a vital role in stabilising voltage quality and power delivery have been considered for many years as an important part of so called microgrids that could be a method to secure the energy supply in power grids with highly distributed feed-ins [3–5]. Moreover, as electric vehicles will have spread extensively, their battery packs are considered to be used to buffer peaks in power demand or dips in power supply [6] in vehicle to grid systems thus demanding sophisticated charging and discharging systems that are based on various power electronic circuits [7–9]. In summary, every battery in battery energy systems is in some way connected to power electronics, especially DC/DC-converters. They are based on controlling semiconducting, transistor-based high frequency switches to modulate the desired voltage or current respectively. Therefore, if costly and bulky filtering measures were not installed, the cells would be under severe high frequency switching stress. High frequency current waves, called 'current ripple', would be induced on the batteries and could lead to battery ageing mechanisms that are still unknown or render known ageing mechanisms more severe. If the results showed no further influence and therefore implied that current ripple does not shorten the battery life, the results could still lead to improvements in the design of battery energy systems as filters that are normally used to even out ripple, e.g., in [9–11], could be reduced in size or even be omitted. It is expected that the severity of current ripple rises with future converters that are based on silicon carbide (SiC) that operate at very high frequencies of several tens or even hundreds of kilohertz [12]. However, the common usage of switching frequencies considerably higher than a few kilohertz are not expected to excite electrochemical reactions thus high frequency current ripple is not expected to have any severe impact on battery ageing just as it is suggested by the following overview.

In the last years, a small amount of research groups have dealt with those particular questions in various ways: In [13], De Breucker at al. investigate the influence of current ripple, induced by a DC/DC-converter connected to a high voltage battery pack that could occur in a plug-in hybrid vehicle. Based on the switching frequency of 8 kHz the double layer is supposed to be mainly influenced by the ripple. Instead, their results show that the temperature plays a much more important role. It should be noted, that only two battery packs have been investigated. In comparison, the authors of [14] superimpose sinusoidal current waves with different frequencies up to 14.8 kHz on the DC current that charges and discharges 15 different 18,650 lithium-ion cells with a nickel cobalt aluminium oxide positive electrode (NCA) in a more detailed ageing test. It is argued that the aforementioned heat generation, leading to faster battery ageing, might also be caused by the imposed ripple current. Yet, a direct formula or factor is not given. A second study that deals with the effects of superimposed sinusoidal current waves can be found in [15]. Instead of NCA, more than 18 commercial cells with a positive electrode that is made of nickel manganese cobalt oxide (NMC) are used. Again, the authors stress that elevated temperatures cover possible effects originating from small current ripple. Finally, Bessman et al. give a short but comprehensive review of additional research in this particular field in [16]. Moreover, they carry out measurements with underlying triangular waves using twelve larger prismatic cells that are more comparable to cells that are used in high energy and/or high power battery storage systems. Compared to the other studies and most of those further mentioned within them, the work that is presented in [14] stands out since it is one of the very few that suggests a significant connection between current ripple and accelerated ageing. Moreover, most explanations remain vague and have to be categorised only as well thought-out assumptions. Thus, this present study has two goals: On the one hand, it is a further investigation on the effects of current ripple induced by practical DC/DC-converters on cylindrical 18,650 NMC lithium-ion batteries. The results of comparative cyclic ageing tests using ripple currents and comparatively unperturbed DC-current respectively, are shown and analysed thoroughly. On the other hand, the authors try to give further explanations as to why it turns out to be quite challenging to find reliable answers to the specific question of how exactly alternating currents affect battery ageing. The particular approach at hand is organised as follows: Section 2 is a detailed description about the hardware setup, the design of the cyclic ageing tests and the investigated battery are introduced. For the ageing tests, a DC/DC-converter to cycle the batteries has been developed. Its working principle and control algorithms are introduced shortly. Furthermore, regarding the test procedure, the converter is accompanied by conventional battery test equipment, that delivers mostly undistorted DC-currents, so that correspondent ageing tests with DC-currents that do not have significant current ripple can be carried out. Those ageing tests follow a set of rules that are established to be able to analyse the preassigned ageing parameters, to make the ageing tests as comparable as possible and to narrow down side effects that could overlay the influence of current ripple. In Section 3 the ageing results for a representative batch of ageing parameters is given, described in detail and more uncommon methods are introduced shortly. The depiction of the ageing results is focused on the comparison between the conventionally aged cells and those aged with ripple current. Subsequently, the results are discussed and analysed critically at the ends of each subsection. To conclude, in Section 4 the results are classified and rated. Based on this, further advice for the practical use of batteries that are connected to power electronics is given so that future battery energy storage systems can be used in an optimal way.

2. Experimental

In this chapter, the circuit for the cyclic ageing tests is proposed, the setup is shown and the design of the ageing tests is elaborated. The battery that has been used in every ageing test is a commercially available 'LG 18650 HE4' cylindrical 18,650 lithium-ion battery with a graphite anode and a cathode made out of cobalt-, nickel- and manganese oxide (NMC). Its nominal capacity is $C_N = 2.5\,\mathrm{A\,h}$, measured at a 1 C discharge from the end-of-charge voltage $U_{EOC} = 4.2\,\mathrm{V}$ to the end-of-discharge voltage $U_{EOD} = 2.5\,\mathrm{V}$.

2.1. Ripple Current Test Circuit

The short literature overview in the introduction has already been a glimpse of how challenging the investigation of the influence of current ripple on battery ageing appears to be. First of all, well suited test equipment has to be found. Keeping the focus on the selected literature, the authors of [14–16] use signal generators to induce sinusoidal or triangular waves on the DC-current flowing in and out of the batteries. This is a very viable approach since the spectra only consist of the fundamental wave in case of the sinusoidal excitation or in the well known form of odd harmonics that decline in orders of $1/n^2$ in case of triangular ripple currents. Thus, these methods have the highest potential for reproducibility but cannot be seen as a practical approach. Especially a pure sinusoidal excitation should be considered carefully as a recent study [17], has shown, that using higher harmonics is a feasible tool to analyse the electrode reactions. Thus, an impact of higher current harmonics on the quality of the electrode reactions cannot be excluded. At the cost of higher noise and a slight signal dependency on the state of the circuit, actual DC/DC-converters might also be used as cycling circuits that induce ripple currents on the batteries as De Breucker et al. have done in [14]. They use a half-bridge converter with a high voltage battery pack. Although one can expect the most practical results, it comes with a lot of constraints such as expected noise because of the voluminous setup and very high currents or voltages, respectively. Besides, a high voltage system is more hazardous and makes repeated tests with more cells or battery packs quite costly.

In this paper, a reasonable compromise is found: Using the widely known half bridge converter as shown for example in [18] at low voltage as a foundation to cycle single cells is still a practical approach that is cheap, scalable and does not have to deal with high voltage and severe noise. Figure 1 shows the basic working principle of the circuit. The battery is located at the low side and is connected directly to the smoothing inductance. At the high-side a 12 V voltage source or a resistive load with a smoothing capacitor are connected to the circuit with a simple switch, depending on whether the the battery is charged or discharged, respectively. Low- and high-side are interconnected with field-effect transistors (FET) and their internal body diodes. In Figure 1a,b, the current flow is given, depending on the operation mode. Assuming lossless and ideally fast switches, an ideal inductor, that U_{CHA} is a constant voltage source, that U_{bat} and u_{C_a} are approximately constant or rather their time constants are large compared to the switching frequency the differential equation

$$-L\frac{di_{bat}}{dt} = u_L = \underbrace{U_{bat} - uU_{CHA}}_{\text{charge mode}} \text{ or } \underbrace{(U_{bat} - uu_{C_a})}_{\text{discharge mode}} \tag{1}$$

with $u \in \{0,1\}$ yields the linearly rising and falling current wave, that is depicted in Figure 1c,d respectively. Thus, the battery current is basically a triangular wave that is induced on a mean value $\bar{i}_{bat}$ comparable with the approach in [16].

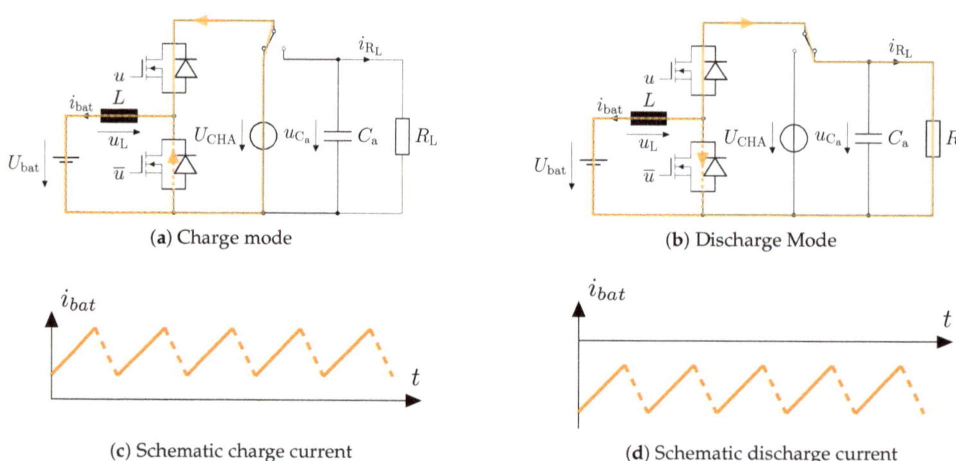

(**a**) Charge mode

(**b**) Discharge Mode

(**c**) Schematic charge current

(**d**) Schematic discharge current

Figure 1. Equivalent circuit diagrams of the half bridge converter used for current ripple ageing tests. The solid line depicts the current flow through the battery while the upper transistor is turned on whereas the dashed line depicts current flow through the battery with the lower transistor being turned on.

Combining the differential equation

$$C_\mathrm{a}\frac{du_{C_\mathrm{a}}}{dt} = \frac{1}{R_\mathrm{L}}u_{C_\mathrm{a}} \tag{2}$$

that affects the system while discharging the battery with Equation (1) yields the full dynamic description of the circuit model in the upper half of Figure 1. However, it is more convenient to separate the equations as indicated by the switch on top of the circuit diagram and design independent control algorithms for charging and discharging, i.e., for the highlighted paths in Figure 1a,b respectively. In both cases, the output variable is the same as the state variable i_bat. The control loop is based on a state feedback controller taken from a textbook such as [19]. It is designed so that the output asymptotically follows the desired value $\bar{i}_\mathrm{bat}$, i.e., the charge or discharge current of the battery. As the other state variable u_{C_a}, that does not appear while charging, is not measured while discharging, a state observer, e.g., as in [19], has to be added. A full block diagram is shown in Figure 2a. As indicated in the picture, the control algorithms are carried out on a microcontroller. In this case, it is an 'ARM Cortex-M4' on an 'Infineon XMC4700 Relax Kit'. Moreover, an interface is implemented so that the cycling circuit is able to communicate with a PC via USB. The incoming measurements on the PC are gathered, visualised in real time and continuously saved with a GUI that is implemented in LabView. On top of that, the case comes with basic I/O-features and a status display. A representative photo of the circuit is shown in Figure 2b.

2.2. Ageing Test Structure

In order to be able to relate ageing phenomena with ripple current, the ageing tests are based on comparison with ageing results of identical ageing tests, that are conducted as a reference test on the conventional test system 'Arbin BTS2000'. It is not the aim of this work to formulate a comprehensive ageing model for the tested battery. That is why the ageing matrix, depicted in Table 1, only consists of two different cycle depths $\Delta DOD_1 = 100\%$ and $\Delta DOD_2 = 10\%$. Moreover, every cell that has been exposed to ripple current is marked with the letter R in a dark yellow or light orange, respectively. Consequently, the cells that are aged by the conventional tests system are marked with a dark green C. ΔDOD_1 has been chosen as one cycle depth so that the ripple current might cause higher ageing rates because of higher or respectively lower high-frequency overpotentials, that shortly deep discharge or overcharge the battery respectively and might remain undetected by a battery management system in a practical application. As it is presented in [20], very low or very high SOC affect rapid SEI-growth, the occurence of Inhomogeneities, co-intercalation of solvents or accelerated decomposition in general. Thus, the main question to be answered is: Is it possible and if so, how is it possible to separate ageing effects and connect them solely to the current ripple? Besides, batteries that are aged with deep full cycles age much faster, see e.g., [21], so that the experiment becomes statistically more evaluable faster due to the possibility to test larger amounts of cells at the same time. In contrast, it is commonly accepted by now that low cycle depths such as ΔDOD_2 around a low or medium $\overline{SOC}$ result in much slower ageing rates. Thus, superimposed ageing mechanisms such as the current ripple appear much stronger so that it should be easier to detect its expected influence. This mindset draws through the other parameters of the ageing tests as well. All tests are conducted in a temperature controlled environment in an oven (Memmert UF55) at an only slightly elevated temperature of 35 °C so that calendar ageing is reduced to a minimum as indicated in [22]. Besides, the mean state of charge is always $\overline{SOC} = 50\%$ and the cells are always cycled with a current rate of $I_{cyc} = 2.5\,\text{A}$ which translates to a current rate of 1 C compared to the nominal capacity of $C_N = 2.5\,\text{A h}$. At this current rate and a switching frequency of $f_S = 5\,\text{kHz}$. the amplitude of the current ripple is going to be around $0.29 I_{cyc}$ at $\overline{SOC} = 50\%$, ranging from $0.22 I_{cyc}$ at the discharge cut-off voltage of 2.5 V to $0.32 I_{cyc}$ at the charge cut-off voltage of 4.2 V, dependant on the battery's terminal voltage.

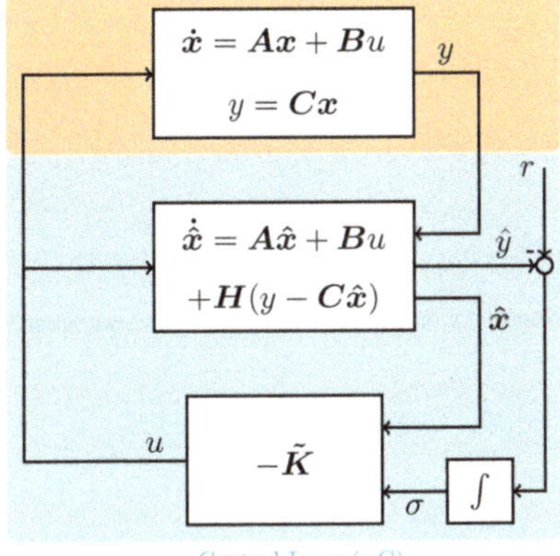

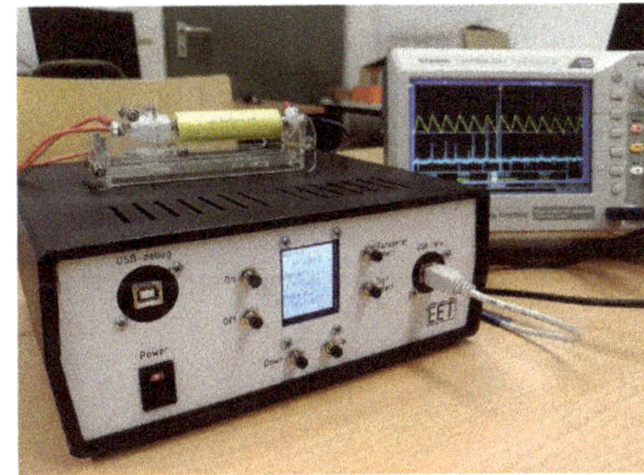

(**a**) State feedback controller with state observer for discharge mode

(**b**) Illustrative photo of the half bridge converter cycling circuit

Figure 2. Observer based control is used since u_{C_a} is not measured. In charging mode, no observer is needed as the only state variable $i_L = -i_{bat}$ is directly measured. In the background of the photo on the right, the ripple current is observable on the oscilloscope.

Table 1. Distribution of the cells between the four different ageing tests. $\overline{SOC}$ and ϑ are kept constant at 50% and 35 °C, respectively.

	Test Systems	
ΔDOD	Conventional	Ripple Current
100%	C_1, C_2, C_3, C_4	$R_1, R_2, R_3, R_4, R_5, R_6$
10 %	C_5, C_6	R_7, R_8

In Figure 3a, the general structure of every ageing test is depicted. It mainly consists of three parts: The cycling with one hundred full cycles at a time, periodical check-up measurements to obtain updated ageing parameters, cycling tests and occasional electrochemical impedance spectroscopy (EIS) to gather information on the dynamic behaviour of the tested battery. It should be further noted that for C_1 and R_1 check up tests have been done every fifty full cycles to have some information about the influence of the check up tests on battery ageing and about the probability of missed information if the check-ups are to far apart from each other. Just as the reference ageing tests, the check-up tests for the rippled batteries are also conducted at the conventional cell tester to minimize differences induced by different test equipment. The general structure and course of the check-up is depicted in Figure 3b. Within the tests, the remaining capacity after a full discharge with 1 C= 2.5 A is measured at the beginning. Afterwards, the cell is fully discharged and charged with a much smaller current, i.e., $|I| = 0.2$ C= 0.5 A, to obtain information that can be used to calculate the differential voltage analysis (DVA) [23]. The rest of the check-up consists of partial discharges and charges and subsequent relaxations over thirty minutes so that the evolution of the cell's inner resistance is also taken care of, albeit not shown in this article as the results do not contribute anything unique to the evaluation. If the cell is still functional, it will be cycled again. In this work, an arbitrary limit such as eighty percent remaining capacity is not used as it is unclear, whether there is any severe ageing because of current ripple, that appears in the nonlinear part of battery ageing [24]. Therefore, ageing tests are not necessarily ended at typical limits.

In addition to that, the dynamic impedance is measured with a Zahner IM6ex every two hundred cycles at three different states of charge, i.e., $SOC_1 = 80\%$, $SOC_2 = 50\%$, $SOC_3 = 20\%$ with a bandwidth of $f \in [100\,\text{mHz}\ 100\,\text{kHz}]$.

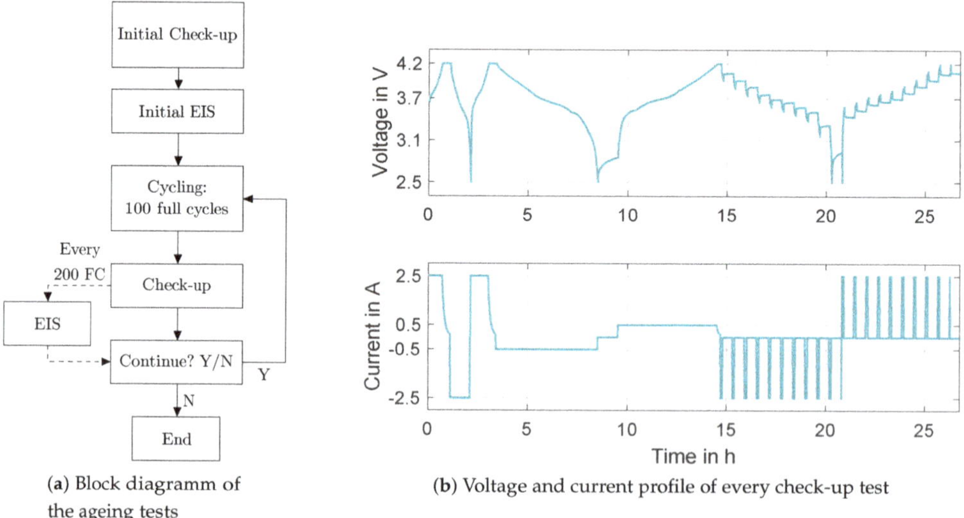

(a) Block diagramm of the ageing tests

(b) Voltage and current profile of every check-up test

Figure 3. Procedure of ageing tests with the current and voltage profile shown on the right side.

3. Results and Discussion

The following section deals with the comparison of the capacity loss, the differential voltage analysis (DVA), the impedance and the distribution of relaxation times (DRT) between the cells, that are either exposed to ripple current or the cells that are cycled with the conventional test system. For each method, the battery groups, related by cycle depth, are compared to each other. Normally, it is refrained from investigating each individual cell as the amount of different potentially affecting parameters renders such assumptions pointless. It is assumed that only a possibly resulting general trend leads to representative and reproducible results.

3.1. Capacity Loss

The capacity diminution is seen as one of the most important parameters to define ageing as it is directly connected to electric vehicle range or the amount of time a storage system can operate. Accordingly, the capacity fade is also referred to as the 'state of health' (SOH) such as in e.g., [20,25] so that the remaining capacity is directly linked to the remaining usability of the battery in practical applications. In Figure 4, the capacity evolution for each tested cell is depicted. The Figure 4a represents the cells that have been cycled with $\Delta DOD_1 = 100\%$ whereas Figure 4b shows the evolution of the capacity for the cells with a much lower ΔDOD_2 of 10%. To reduce the business of Figure 4a due to the higher amount of tested cells, the same measurements are shown in Figure 5 as mean values with errorbars, that represent the upper and lower bounds of confidence intervals for a probability of 95%. It should be noted that the markers in Figure 5 are arbitrarily set at every fifty equivalent full cycles as the results have been interpolated to make the statistical analysis possible. Besides, the suddenly increasing slope of the ripple current graph in Figure 5 comes from premature cell failure of R2, R3 and R5 as indicated in Figure 4a. Throughout all tests a major similarity can be noted: The capacity drops quite steeply in the beginning of the tests. Interestingly, the capacity depletion of the cells cycled with ΔDOD_1 slows down at roughly $0.8C_0$ which is a common reference point for the transition between linear and nonlinear ageing [24,26]. Later on, the typical spread between faster and severely faster aged cells as reported in [27] is observable. Furthermore, it becomes visible in the last part of the graph that the rippled cells and the conventionally aged cells are grouped respectively which is easier to distinguish in Figure 5. At the same time, the upper and lower error bounds become much larger. However, this grouping should not be over-interpreted as it only occurs at late stages of battery life beyond a *SOH* of approximately 70 % and is expected to be linked to premature cell failures (see above) and the volatile region of nonlinear battery ageing that could also be due to statistical spread induced by production tolerance as investigated in [27].

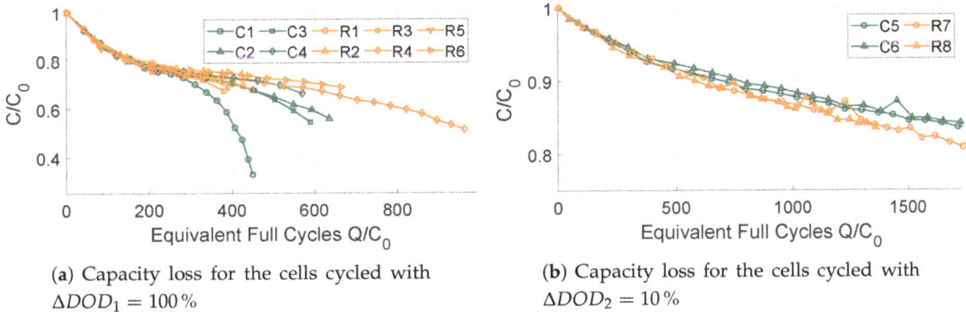

(a) Capacity loss for the cells cycled with $\Delta DOD_1 = 100\%$

(b) Capacity loss for the cells cycled with $\Delta DOD_2 = 10\%$

Figure 4. Results of the capacity loss for cells cycled with high and low $\Delta DOD_{1/2}$. Instead of $C_N = 2.5\,\text{A h}$ each individual starting capacity C_0 has been used as they differ slightly from C_N.

The aforementioned grouping is also visible for ΔDOD_2, as seen in Figure 4b. In opposition to the fully cycled cells, R7 and R8 age a bit faster than the conventionally aged reference cells C5 and C6. Moreover, the contradictory observation between Figures 4b and 5 cannot be explained satisfactorily in another way than statistical spread. It is assumed, that an absence of the unexpected early failure of cells R2, R3 and R5 would have led to a more similar trend of the mean values. Moreover, the uncertainty of the statistical approach rises as the number of analysed cells diminish. Therefore, the analysis of the capacity loss leads to the assumption that the effect of a deep cycle depth greatly outmatches the influence of current ripple whereas the influence of current ripple could be visible for the partially cycled cells instead.

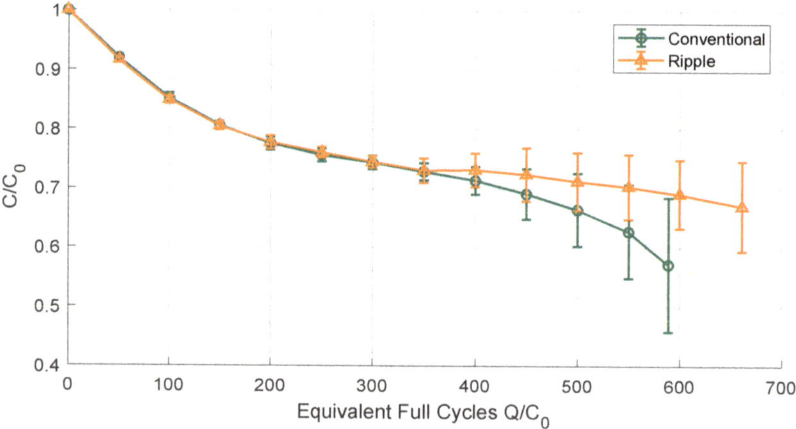

Figure 5. Capacity loss of cells cycled with full cycles as mean values with a confidence interval of 5%.

3.2. Differential Voltage Analysis (DVA)

In recent years, the differential voltage analysis (DVA) has become a common tool to analyse the behaviour of the electrodes of a lithium-ion battery, primarily the behaviour of the anode [23,28]. It is based on the derivation of the voltage with respect to the transferred charge or the respective SOC. Characteristic local maxima or minima, respectively, correlate with the phase changes of the graphite that is commonly used as an anode material such as within the investigated cell. Thus, the ageing effects are visible in the depiction of the DVA-curves. However, the phase changes are only clearly visible for very low charge or discharge currents that do not cause high and therefore overlapping overpotentials. In this study, a current of $I_{DVA} = 0.5\,A$ which translates to $0.2C_0$ is used as a compromise between visibility of characteristic elements and measurement time.

In Figure 6, three exemplary graphics are shown to illustrate the results of the DVA. The pictures are separated by $\Delta DOD_{1/2}$ and the respective *SOH*, as on the left side in Figure 6a, the cells cycled with ΔDOD_1 are shown after a capacity loss of roughly 20 % whereas on the right side in Figure 6b the cells cycled with ΔDOD_1 are shown by the time the cells approximately reached a capacity loss of 30 %. As it can be seen in Figure 4a, the cells are not necessarily cycled with the same amount of cycles at that particular point. Besides, one cell is left out in Figure 6b. The rippled cell R2 has broken down before it has reached the desired capacity drop of 30 %, so that it is not shown in this particular picture. For further comparison, the DVA-curves, obtained at the initial check-up are also shown in light grey for each cell. They overlap each other well which is an indicator for the highly precise and reliable manufacturing of the cells. As it can be seen in Figure 6a,b, the overlapping continues as the cells age. This is expected until the cells reach a *SOH* of 80 % and is even maintained in regions beyond a typical capacity loss of 20 %. As long as the DVA is executed for cells with the same capacity loss, even a distinction between rippled cells and conventionally aged cells is impossible let alone cells of the same group. In Figure 6c, it is possible to spot the different curves. However, the grouping that has been reported for the capacity loss is not as clearly visible as before. As a reason of the aforementioned very minor differences, especially for cells cycled with ΔDOD_1, the only further analysis the plots could be used for, is the analysis of cyclic ageing effects in general as the diminution of the local extrema due to degradation of the anode and shortening due to capacity loss are typical features. However, the authors refrain from this as it is not in the scope of this work and discussed thoroughly in the literature [23,28,29]. In addition to that, a look at Figure 6 in accordance with [29] reveals that, given the same *SOH*, i.e., the same capacity loss, substantial changes of the anode surface and therefore the voltage plateaus of its intercalation reaction are mostly linked to the cycle depth. This is one reason why the *SOH* has been used as the control variable to pick the proper check-up test for comparison.

Thus, current ripple does not seem to contribute in a way the cycle depth already does on its own to the loss of active material or loss of active lithium , respectively as it would have been observable in the DVA-curves otherwise.

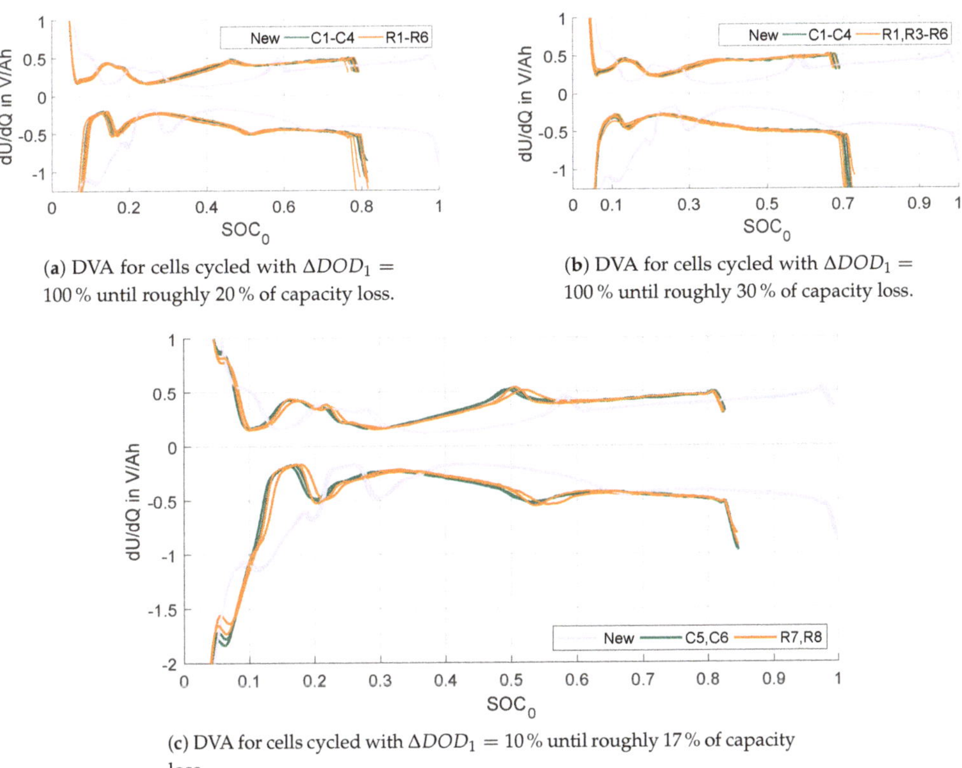

(**a**) DVA for cells cycled with $\Delta DOD_1 = 100\,\%$ until roughly 20 % of capacity loss.

(**b**) DVA for cells cycled with $\Delta DOD_1 = 100\,\%$ until roughly 30 % of capacity loss.

(**c**) DVA for cells cycled with $\Delta DOD_1 = 10\,\%$ until roughly 17 % of capacity loss.

Figure 6. Comparison of the DVA of aged batteries for $\Delta DOD_1 = 100\%$ and $\Delta DOD_2 = 10\%$. The initial DVA-curves of every cell are shown in light grey.

3.3. Impedance Measurements

For a lot of years, the electrochemical impedance spectroscopy (EIS) has been one of the most widely used methods to analyse any electrochemical system [30]. Based on the assumption that electrochemical systems are approximately linear and time-invariant (LTI), a galvanostatic excitation with a sinusoidal current over a broad band of frequencies ω yields the impedance

$$\underline{Z}(\omega) = \frac{\underline{U}(\omega)}{\underline{I}(\omega)}.$$

This impedance can be used to derive a dynamical model of the measured cell or to analyse ageing effects as the behaviour of the impedance is directly linked to corresponding characteristics of the cell such as the electrode reactions or diffusion. The trend of the impedance is normally visualised as a Nyquist plot in the complex plane. This visualisation can be seen in Figure 7a,b in the same way as the results for the DVA are presented in Figure 6, i.e., at the same SOH to minimize unwanted discrepancies between the groups of cells because of different capacity losses. A picture that shows the spectra at $SOH = 80\,\%$ is omitted as it does not supply any further information compared to the impedance spectra shown at a capacity loss of 30 %. The shown frequency range is $f_{\text{EIS}} \in [1\,\text{mHz}\ 10\,\text{kHz}]$ since high frequency parts above $10\,\text{kHz}$

do not show any distinguishable differences between the aged cells. Moreover, the inductive reactance rises significantly at higher frequencies so that this area has been also cut due to better visibility of the capacitive area. Furthermore, general statements about the alteration of the spectrum because of cyclic ageing are neglected again to keep the focus on the comparison. As expected, the intersection of the real axis, often referred to as the inner resistance R_i, e.g., in [30], rises as the cells age. It should be noted that the spread between the initial intersections has been equalised in Figure 7a,b to improve comparability. Taking this into account, a further grouping is visible as most inner resistances of the rippled cells have grown slightly larger compared to the conventionally aged cells. More information is gathered, if the polarisation of the electrodes is taken into account. Considering the new cells, the representation of the electrodes cannot be distinguished. This changes as the batteries age since both flattened semicircles separate from each other and become wider. Except for one rippled cell in each group, i.e., R6 and R8, no clear differentiation between the rippled and the conventionally aged cells is observed. Moreover, it is challenging to derive the most prominent time constants that cause the spectra at hand, yet this valuable information is a nominal asset to compare rippled and non-rippled cells even further.

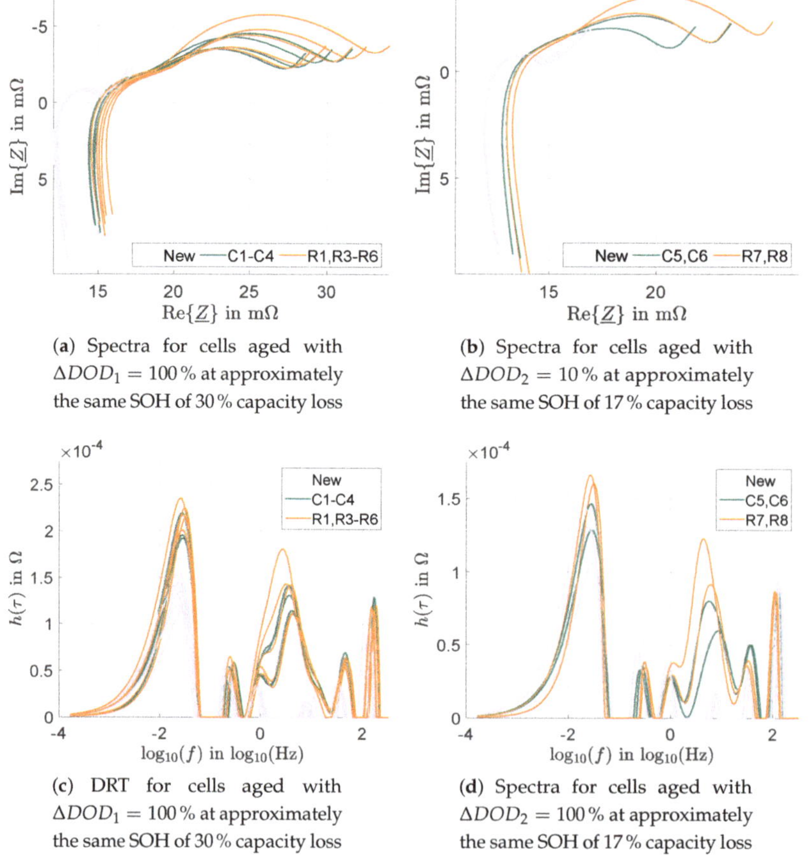

(**a**) Spectra for cells aged with $\Delta DOD_1 = 100\,\%$ at approximately the same SOH of 30 % capacity loss

(**b**) Spectra for cells aged with $\Delta DOD_2 = 10\,\%$ at approximately the same SOH of 17 % capacity loss

(**c**) DRT for cells aged with $\Delta DOD_1 = 100\,\%$ at approximately the same SOH of 30 % capacity loss

(**d**) Spectra for cells aged with $\Delta DOD_2 = 100\,\%$ at approximately the same SOH of 17 % capacity loss

Figure 7. Evaluation of impedance measurements as Nyquist curves and as plots of the DRT. The variation of the intersection of the real axis is corrected in the upper pictures whereas the resistance R_i is also substracted from the spectra to get the DRT measurements in accordance to [31,32]. The initial curves are shown in light grey respectively.

To achieve this goal and to validate whether the slight outlier mentioned before is also visible in other ways, the calculation of the distribution of relaxation times (DRT) has proven to be a suitable tool to analyse impedance spectra further [33]. As explained in [30], the polarisation of dielectric materials such as electrolytes can neither be fully described by a single time constant nor a single RC-circuit, respectively but rather as a distribution of time constants that is often referred to as the distribution of relaxation times. It is explained by the representation of the impedance as

$$\underline{Z}(\omega) = R_0 + R_{\text{pol}} \int_0^\infty \frac{g(\tau)}{1 + j\omega\tau} \mathrm{d}\tau \tag{3}$$

with the inner resistance R_0 and the distribution function $g(\tau)$ that is usually normalised by

$$\int_0^\infty g(\tau)\, \mathrm{d}\tau = 1$$

so that the polarisation resistance R_{pol}, that represents the width of the semicircle in typical impedance spectra of batteries is separated from the distribution. Thus, $g(\tau)$ needs to be calculated. Considering measurements with a limited amount of m data points $\underline{Z}(\omega)$ over a limited set of m excitation frequencies ω and an arbitrarily chosen amount of n time constants τ_k, the integral becomes the discrete sum

$$\underline{Z}(\omega) = R_0 + R_{\text{pol}} \sum_{k=1}^{n} \frac{g(\tau_k)}{1 + j\omega\tau_k}. \tag{4}$$

As mentioned in [34], this task requires the calculation of an ill posed problem because the improper 'Fredholm integral' in Equation (3) or the corresponding sum in Equation (4) has to be calculated. According to [31,32,34] a promising approach is the 'Tikhonov-regularisation' that converts Equation (3) to the optimisation problem

$$\min \left\{ ||\mathbf{A}\mathbf{x} - \mathbf{b}||^2 + ||\lambda\mathbf{x}||^2 \right\}.$$

It consists of the matrix $\mathbf{A} \in \mathbb{R}^{m \times n}$, representing the unweighted RC-elements with the arbitrarily chosen time constants τ_k, whose quantity and bandwidth should surpass those of the angular frequencies ω_i of the measurement [32], the vector $\mathbf{b} \in \mathbb{R}^m$, representing the measured impedance $\underline{Z}(\omega_i)$ and the optimisation factor λ that has to be chosen carefully, [31,34]. The distribution function $g(\tau)$ is stored in $\mathbf{x}$ after a successful numerical optimisation with a feasible solving method such as the non-negative least squares (NNLS) algorithm [31]. Much more detailed information on the calculation of the DRT is found in [31], too.

In the lower part of Figure 7 the result of the DRT-analysis is shown in correspondence to the spectra in the upper part. Thus, the graphs basically show the same measurement. However, the prominently contributing time constants are clearly visible in Figure 7c,d. Besides, it should be noted that the bandwidth shown in these depictions corresponds to the a priori chosen time constants τ_k and ranges from 0.1 mHz to roughly 300 Hz to get the most meaningful representation. A broader bandwidth of the time constants up to several MHz is used to extend the DRT to the inductive branch in adaption to [31], modelled as a distribution of RL-circuits which are proposed in [35]. However, these results are not shown as they do not provide any further useful information. Instead of $g(\tau)$,

$$h(\tau) = R_{\text{pol}}g(\tau)$$

is shown on the vertical axis. In both figures, five major peaks can be detected that are in good accordance with the literature, for example [30] or [33]. The first peak from left to right represents the diffusion branch. As in the spectra, no distinctive difference between rippled cells and conventionally aged ones is found. In general, the rise of the peak over the battery ageing implies a flattening of the diffusion branch that is not detected without any further analysis in the spectra. The following peaks are directly connected to the polarisation of the electrodes with two smaller peaks that most likely represent the polarisation at the cathode and the larger peak in the middle corresponding to the main anode reaction. These peaks mainly shape the capacitive parts of the spectra. The last peak is connected to the interfaces between the current collectors and the active mass. Generally, the observations considering the spectra can be validated by the DRT. Again, cells R6 and R8 are prominent. Their middle peaks are clearly elevated as compared to the other cells.

Another approach to visualise the information given in the impedance data is shown in Figure 8. In this picture, the normalised height of the largest peak in the middle of the DRT, representing the polarisation of the anode, is plotted against the capacity loss as shown in Figure 4a for cells cycled with ΔDOD_1. A slight grouping just as in Figure 4a is detectable as an addition to the general observation that the polarisation of the anode starts to get worse more rapidly as the capacity deterioration is getting slower. In this picture, R5 is not shown because the initial EIS-measurement at zero cycles is missing so that a relative examination is not possible.

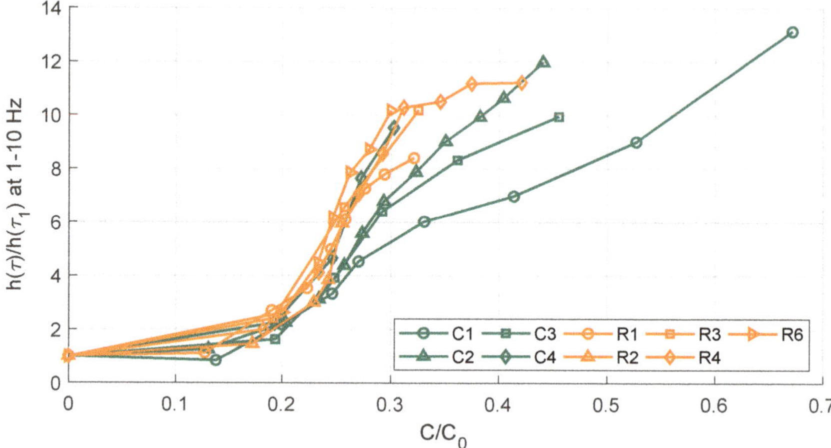

Figure 8. Trend of the anode peak at roughly 10 Hz in the DRT-measurements over capacity loss for cells cycled with $\Delta DOD_1 = 100\%$.

As expected, the polarisation of the electrodes changes over time as the cell reactions are constrained more and more which leads to a rising real part at lower frequencies as the batteries age. In [20], most of these changes in the dynamic behaviour of the electrodes are linked to SEI-growth which is most prominently accelerated by a high SOC and higher temperatures. Thus, as the rather small amplitude of the current ripple does not lead to periodic overcharges due to transient higher overpotentials that are not recognized by the test circuit, different temperatures while cycling should lead to different ageing curves. Given the well known fact, that the temperature is linked to ohmic losses calculated by multiplying the square of the root mean square value (hereafter: RMS-value) of the current with the internal resistance, a different RMS-value should lead to a different temperature because the RMS-value represents the equivalent DC-value of the alternating current that would convert the same amount of energy at a resistive load which is solely heat. The RMS-value of a purely direct current is the same as the mean value that is used to charge and discharge the batteries whereas the RMS-value of a triangular wave as in Figure 1 is calculated by

$$I_{\mathrm{RMS,R}} = I_{\mathrm{min,R}} + \frac{1}{\sqrt{3}}\Delta I_{\mathrm{R}}$$

so that it is higher than the mean value which could lead to higher temperatures compared to an undistorted direct current. However, further measurements have shown that the difference in surface temperature of the cells between rippled and conventionally aged cells is below 2 K. This fits the observation that the ripple current does not have any clear influence on the dynamic battery behaviour that exceeds the influence of cyclic ageing in general. The advantage of the DRT that the most influential processes at the electrodes are visible separately further supports the aforementioned claim as no clear deviation between the cell groups is visible for any kind of diffusion or reaction process. Moreover, no direct connection between ripple current and the outliers, visible in Figure 7a–d, respectively could be found. It is expected that these cells represent the unpredictable spread that occurs in later parts of battery ageing [27]. To conclude, the spectra are used as a further possible explanation as to why the ripple does not have any severe impact. The assumption, e.g., in [14,16], that the excitation frequency of the current ripple, i.e., the switching frequency of the DC/DC-converter, has the highest impact on the dynamic behaviour of the battery if it is still in its capacitive range and thus provoking unwanted reactions at the electrodes could be another explanation for the lack of impact of high frequency current ripple on battery ageing. Switching frequencies of practical DC/DC-converters are typically located at several kilohertz which usually corresponds to the inductive branch of the battery as illustrated in Figure 9. In these drawings, the trends over ageing of the impedance spectra of two arbitrarily chosen batteries C2 and R4 with and without ripple current are depicted. Moreover, an excitation frequency of 5 kHz that is also the switching frequency of the converter is marked for each spectrum. Neither does the mark change position significantly nor does the inductive branch show any clear alteration because of ageing. As stated by [35,36], the inductive branch is mostly affected by the geometry of the cell and the current collectors which does not change due to cyclic ageing not to mention current ripple.

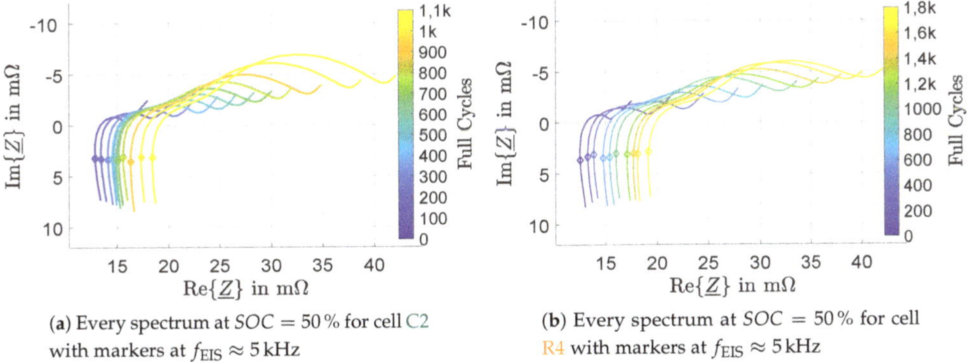

(a) Every spectrum at $SOC = 50\,\%$ for cell C2 with markers at $f_{\text{EIS}} \approx 5\,\text{kHz}$

(b) Every spectrum at $SOC = 50\,\%$ for cell R4 with markers at $f_{\text{EIS}} \approx 5\,\text{kHz}$

Figure 9. Impedance spectra over ageing for a conventionally aged cell and a rippled one at $SOC = 50\,\%$ to illustrate the position of the converter's switching frequency in the spectra.

4. Conclusions

In this article, the influence of practical DC/DC-converter induced high frequency current ripple on the ageing of conventional 18650 lithium-ion batteries is investigated. After numerous ageing tests, the measurements strongly indicate that there is no significant effect of half bridge converter induced current ripple on battery ageing that outmatches the influence of a deep DOD. However, by reducing the influence of the cycle depth by using $\Delta DOD_2 = 10\,\%$, slight deviations between conventionally aged and rippled cells are visible, regardless of the measurement as seen in Figures 4b, 6c and 7b,d. Yet, a distinct relation between current ripple and these deviations remains debatable. These statements incorporate themselves into the literature and support the findings of [16] and the literature overview within. In addition to further literature, e.g., [13–15] the research at hand is able to provide some more attempts to explain whether current ripple could be harmful in general and why the detection of any current ripple affected ageing parameters is rather difficult.

Investigating the influence of current ripple or, more generally speaking, the influence of AC-excitation on battery ageing implies cyclic ageing tests that most likely influence battery ageing more than alternating currents as long as the ageing test is made from a practical perspective. The amplitude of the current should stay within the battery specifications and the current profile itself is orientated to reproduce typical practical dimensions, i.e., limited current ripple. In concordance with [14,16], it is expected that alternating currents with amplitudes beyond battery specifications or at least designed in a way so that the effective value is drastically higher than the mean value lead to elevated heating of the batteries. However, overextending this approach would lead to accelerated ageing because of higher temperatures resulting from the high rms-value of the ripple currents. For comparison: In this study, the temperatures of the rippled cells are elevated by not more than 2 K compared to the conventionally cycled cells. It is expected that ageing tests that overextend on the deviation of the effective value from the mean value of the ripple current would reveal that the battery ageing would be influenced much more by the presence of considerably elevated temperatures than the influence of ripple currents. On top of that, the influence of elevated temperatures has already been discussed satisfactorily in the literature, e.g., in [21]. Moreover, such studies would be comparable to the result of this study that a large DOD affects ageing much more than current ripple. In general, indicated by the slight deviations between the cells cycled with $\Delta DOD_2 = 10\,\%$, assumed influences of current ripple are only visible if other ageing influences are minimised.

Further Work

This article joins its solid foundation of referred literature, e.g., [13,15,16], that collectively doubt the influence of current ripple or alternating current on battery ageing. However, generalisation and transferability remain problematic as the vast amount of possible parameters connected to ripple currents that could influence battery ageing produce countless combinations that could be tested. For example, different DC/DC-converters with different current wave of forms such as bidirectional flyback converters or resonant converters could lead to new insights. Besides, changing the switching frequency could lead to different results as indicated in Section 2 in the same way different cell geometries, cell sizes and electrode materials or even electrolytes could lead to alternative behaviour if ripple current is induced. Nonetheless, the likelihood of current ripple not needing to be addressed carefully in practical applications could lead to lower development cost and less filtering effort in future energy storage systems. In [37,38] the component costs of power electronic systems are evaluated. Although the authors do not take batteries into account, their research is appropriate for a rough estimate. According to the studies, a cost reduction up to 20% or even 30% in some cases could be achieved because lighter and more compact filter capacitors and inductors could be used. By adding batteries, these values would be reduced fairly, yet smaller filter capacitors and inductors still most likely lead to cheaper, lighter and more compact energy storage systems which could be a crucial, prospective benefit in highly competitive markets such as the automobile industry.

Author Contributions: Conceptualization, P.K.P.F.; data curation, P.K.P.F.; formal analysis, P.K.P.F.; investigation,P.K.P.F.; methodology, P.K.P.F.; resources, J.K.; software, P.K.P.F.; supervision, J.K.; validation, P.K.P.F.; visualization, P.K.P.F.; writing–original draft preparation, P.K.P.F.; writing–review and editing, J.K.;

Funding: This research received no external funding.

Conflicts of Interest: The authors declare no conflict of interest.

Abbreviations

The following abbreviations are used in this manuscript:

DC	direct current
NCA	Nickel cobalt aluminium oxide
NMC	Nickel cobalt manganese oxide
GUI	Graphical user interface
SOC	State of charge
DOD	Depth of discharge
SOH	Sate of health
SEI	Solid electrolyte interphase
EIS	Electrochemical impedance spectroscopy
DVA	Differential voltage analysis
DRT	Distribution of relaxation times
LTI	Linear and time-invariant
NNLS	Non-negative least squares

References

1. Kittner, N.; Lill, F.; Kammen, D.M. Energy storage deployment and innovation for the clean energy transition. *Nat. Energy* **2017**, *2*, 1–6. [CrossRef]
2. Gür, T.M. Review of electrical energy storage technologies, materials and systems: Challenges and prospects for large-scale grid storage. *Energy Environ. Sci.* **2018**, *11*, 2696–2767. [CrossRef]
3. Lasseter, R.H. MicroGrids. In Proceedings of the 2002 IEEE Power Engineering Society Winter Meeting. Conference Proceedings (Cat. No. 02CH37309), New York, NY, USA, 27–31 January 2002; Voume 1, pp. 305–308. [CrossRef]
4. Gkountaras, A.; Dieckerhoff, S.; Sezi, T. Real Time Simulation and Stability Evaluation of a Medium Voltage Hybrid Microgrid. In Proceedings of the 7th IET International Conference on Power Electronics, Machines and Drives (PEMD 2014), Manchester, UK, 8–10 April 2014; pp. 1–6. [CrossRef]
5. Fang, J.; Tang, Y.; Li, H.; Li, X. A Battery/Ultracapacitor Hybrid Energy Storage System for Implementing the Power Management of Virtual Synchronous Generators. *IEEE Trans. Power Electron.* **2018**, *33*, 2820–2824. [CrossRef]
6. Yoldas, Y.; Önen, A.; Muyeen, S.; V., V.A.; Alan, I. Enhancing smart grid with microgrids: Challenges and opportunities. *Renew. Sustain. Energy Rev.* **2017**, *72*, 205–214. [CrossRef]
7. Gao, F.; Gu, X.; Ma, Z.; Zhang, C. Redistributed Pulse Width Modulation of MMC Battery Energy Storage System under Submodule Fault Condition. *IEEE Trans. Power Electron.* **2019**. [CrossRef]
8. Rivera, S.; Wu, B. Electric Vehicle Charging Station With an Energy Storage Stage for Split-DC Bus Voltage Balancing. *IEEE Trans. Power Electron.* **2017**, *32*, 2376–2386. [CrossRef]
9. Tran, D.H.; Vu, V.B.; Choi, W. Design of a High-Efficiency Wireless Power Transfer System With Intermediate Coils for the On-Board Chargers of Electric Vehicles. *IEEE Trans. Power Electron.* **2018**, *33*, 175–187. [CrossRef]
10. Bala, S.; Tengnér, T.; Rosenfeld, P.; Delince, F. The effect of low frequency current ripple on the performance of a Lithium Iron Phosphate (LFP) battery energy storage system. In Proceedings of the 2012 IEEE Energy Conversion Congress and Exposition (ECCE), Raleigh, NC, USA, 15–20 September 2012; pp. 3485–3492. [CrossRef]
11. Soares, R.; Djekanovic, N.; Wallmark, O.; Loh, P.C. Integration of Magnified Alternating Current in Battery Fast Chargers based on DC-DC Converters using Transformerless Resonant Filter Design. *IEEE Trans. Transp. Electrif.* **2019**. [CrossRef]
12. Bertelshofer, T.; Horff, R.; Maerz, A.; Bakran, M. A performance comparison of a 650 V Si IGBT and SiC MOSFET inverter under automotive conditions. In Proceedings of the PCIM Europe 2016, International Exhibition and Conference for Power Electronics, Intelligent Motion, Renewable Energy and Energy Management, Nuremberg, Germany, 10–12 May 2016; pp. 1–8.

13. De Breucker, S.; Engelen, K.; D'hulst, R.; Driesen, J. Impact of current ripple on Li-ion battery ageing. In Proceedings of the 2013 World Electric Vehicle Symposium and Exhibition (EVS27), Barcelona, Spain, 17–20 November 2013; pp. 1–9. [CrossRef]

14. Uddin, K.; Moore, A.D.; Barai, A.; Marco, J. The effects of high frequency current ripple on electric vehicle battery performance. *Appl. Energy* **2016**, *178*, 142–154. [CrossRef]

15. Brand, M.J.; Hofmann, M.H.; Schuster, S.S.; Keil, P.; Jossen, A. The Influence of Current Ripples on the Lifetime of Lithium-Ion Batteries. *IEEE Trans. Veh. Technol.* **2018**, *67*, 10438–10445. [CrossRef]

16. Bessman, A.; Soares, R.; Wallmark, O.; Svens, P.; Lindbergh, G. Aging effects of AC harmonics on lithium-ion cells. *J. Energy Storage* **2019**, *21*, 741–749. [CrossRef]

17. Harting, N.; Wolff, N.; Röder, F.; Krewer, U. Nonlinear Frequency Response Analysis (NFRA) of Lithium-Ion Batteries. *Electrochim. Acta* **2017**, *248*, 133–139. [CrossRef]

18. Erickson, R.W.; Maksimovic, D. *Fundamentals of Power Electronics*; Kluwer Academic Publishers: Norwell, MA, USA, 2001; Volume 2.

19. Khalil, H.K. *Nonlinear Systems*; Prentice Hall: Upper Saddle River, NJ, USA, 2002; Volume 3.

20. Vetter, J.; Novák, P.; Wagner, M.R.; Veit, C.; Möller, K.C.; Besenhard, J.O.; Winter, M.; Wohlfahrt-Mehrens, M.; Vogler, C.; Hammouche, A. Ageing mechanisms in lithium-ion batteries. *J. Power Sources* **2005**, *147*, 269–281. [CrossRef]

21. Schmalstieg, J.; Käbitz, S.; Ecker, M.; Sauer, D.W. A holistic aging model for Li(NiMnCo)O2 based 18650 lithium-ion batteries. *J. Power Sources* **2014**, *257*, 325–334. [CrossRef]

22. Ecker, M.; Nieto, N.; Käbitz, S.; Schmalstieg, J.; Blanke, H.; Warnecke, A.; Sauer, D.U. Calendar and cycle life study of Li(NiMnCo)O2-based 18,650 lithium-ion batteries. *J. Power Sources* **2014**, *248*, 839–851. [CrossRef]

23. Bloom, I.; Jansen, A.N.; Abraham, D.P.; Knuth, J.; Jones, S.A.; Battaglia, V.S.; Henriksen, G.L. Differential voltage analyses of high-power, lithium-ion cells: 1. Technique and application. *J. Power Sources* **2005**, *139*, 295–303. [CrossRef]

24. Schuster, S.F.; Bach, T.; Fleder, E.; Müller, J.; Brand, M.; Sextl, G.; Jossen, A. Nonlinear aging characteristics of lithium-ion cells under different operational conditions. *J. Energy Storage* **2015**, *1*, 44–53. [CrossRef]

25. Zou, Y.; Hu, X.; Ma, H.; Li, S.E. Combined State of Charge and State of Health estimation over lithium-ion battery cell cycle lifespan for electric vehicles. *J. Power Sources* **2015**, *273*, 793–803. [CrossRef]

26. Rohr, S.; Müller, S.; Baumann, M.; Kerler, M.; Ebert, F.; Kaden, D.; Lienkamp, M. Quantifying Uncertainties in Reusing Lithium-Ion Batteries from Electric Vehicles. *Procedia Manuf.* **2017**, *8*, 603–610. [CrossRef]

27. Baumhöfer, T.; Brühl, M.; Rothgang, S.; Sauer, D.U. Production caused variation in capacity aging trend and correlation to initial cell performance. *J. Power Sources* **2014**, *247*, 332–338. [CrossRef]

28. Petzl, M.; Kasper, M.; Danzer, M.A. Lithium plating in a commercial lithium-ion battery—A low-temperature aging study. *J. Power Sources* **2015**, *275*, 799–807. [CrossRef]

29. Keil, P.; Jossen, A. Calendar Aging of NCA Lithium-Ion Batteries Investigated by Differential Voltage Analysis and Coulomb Tracking. *J. Electrochem. Soc.* **2017**, A6066–A6074. [CrossRef]

30. Barsoukov, E.; Macdonald, J.R. *Impedance Spectroscopy*; Wiley-Interscience: Hoboken, NJ, USA, 2005; Volume 2.

31. Hahn, M.; Schindler, S.; Triebs, L.C.; Danzer, M.A. Optimized Process Parameters for a Reproducible Distribution of Relaxation Times Analysis of Electrochemical Systems. *Batteries* **2019**, *5*, 43. [CrossRef]

32. Danzer, M.A. Generalized Distribution of Relaxation Times Analysis for the Characterization of Impedance Spectra. *Batteries* **2019**, *5*, 53. [CrossRef]

33. Schmidt, J.P.; Chrobak, T.; Ender, M.; Illig, J.; Klotz, D. Ivers-Tiffée, E. Studies on LiFePO4 as cathode material using impedance spectroscopy. *J. Power Sources* **2011**, *196*, 5342–5348. [CrossRef]

34. Ivers-Tiffée, E.; Weber, A. Evaluation of electrochemical impedance spectra by the distribution of relaxation times. *J. Ceram. Soc. Jpn.* **2017**, *125*, 193–201. [CrossRef]

35. Korth Pereira Ferraz, P.; Schmidt, R.; Kober, D.; Kowal, J. A high frequency model for predicting the behavior of lithium-ion batteries connected to fast switching power electronics. *J. Energy Storage* **2018**, *18*, 40–49. [CrossRef]

36. Osswald, P.; Erhard, S.; Noel, A.; Keil, P.; Kindermann, F.; Hoster, H.; Jossen, A. Current density distribution in cylindrical Li-Ion cells during impedance measurements. *J. Power Sources* **2016**, *314*, 93–101. [CrossRef]

37. Burkart, R.; Kolar, J.W. Component cost models for multi-objective optimizations of switched-mode power converters. In Proceedings of the 2013 IEEE Energy Conversion Congress and Exposition, Denver, CO, USA, 15–19 September 2013; pp. 2139–2146. [CrossRef]
38. Wang, H.; Wang, H.; Zhu, G.; Blaabjerg, F. Cost assessment of three power decoupling methods in a single-phase power converter with a reliability-oriented design procedure. In Proceedings of the 2016 IEEE 8th International Power Electronics and Motion Control Conference (IPEMC-ECCE Asia), Hefei, China, 22–26 May 2016; pp. 3818–3825. [CrossRef]

 *sustainability*

Article

Lifetime Analysis of Energy Storage Systems for Sustainable Transportation

Peter Haidl [1,*], Armin Buchroithner [1], Bernhard Schweighofer [1], Michael Bader [2] and Hannes Wegleiter [1]

1   Institute of Electrical Measurement and Measurement Signal Processing, Energy Aware Systems Group, Graz University of Technology, 8010 Graz, Austria; armin.buchroithner@tugraz.at (A.B.); bernhard.schweighofer@tugraz.at (B.S.); wegleiter@tugraz.at (H.W.)
2   Institute of Machine Components and Methods of Development, Graz University of Technology, 8010 Graz, Austria; bader@tugraz.at
*   Correspondence: haidl@tugraz.at

Received: 15 October 2019; Accepted: 23 November 2019; Published: 27 November 2019

Abstract: On the path to a low-carbon future, advancements in energy storage seem to be achieved on a nearly daily basis. However, for the use-case of sustainable transportation, only a handful of technologies can be considered, as these technologies must be reliable, economical, and suitable for transportation applications. This paper describes the characteristics and aging process of two well-established and commercially available technologies, namely Lithium-Ion batteries and supercaps, and one less known system, flywheel energy storage, in the context of public transit buses. Beyond the obvious use-case of onboard energy storage, stationary buffer storage inside the required fast-charging stations for the electric vehicles is also discussed. Calculations and considerations are based on actual zero-emission buses operating in Graz, Austria. The main influencing parameters and effects related to energy storage aging are analyzed in detail. Based on the discussed aging behavior, advantages, disadvantages, and a techno-economic analysis for both use-cases is presented. A final suitability assessment of each energy storage technology concludes the use-case analysis.

Keywords: flywheel energy storage; FESS; e-mobility; battery; supercapacitor; lifetime comparison; charging station; renewable energy storage

---

1. Introduction

Despite the enormous effort put into the reduction of greenhouse gases, CO_2 emissions are still increasing. Road transport contributes up to 25 percent to the CO_2 emissions and represents one of the fastest-growing economic sectors [1]. A possible strategy to reduce local emissions is to increase the share of electric mobility consistently. Observing the technical developments of zero-emission vehicles in recent years, especially energy storage, has proved to be the bottleneck. Despite intensive research activities, mobile energy storage is still the limiting factor, curbing the success of hybrid and electric vehicles.

Since the direct storage of electrical energy can be realized only by the capacitors and coils, indirect storage methods prevail. This means that in a first step, the electrical energy is converted into another form of energy and subsequently stored for later reconversion into electrical energy. In Figure 1, a short classification into mechanical, electrochemical, chemical, electrical and thermal energy storage systems is given.

When energy storage is discussed in the context of sustainable transportation, the first topic that comes to mind is electrochemical batteries for electric vehicles (EVs). Battery electric vehicles—without a doubt—play an important role in our path towards zero-emission mobility, but many experts agree that the energy revolution will require a mix of different energy storage solutions and transportation modes, as the "one-size-fits-all-solution" is yet to be invented [2,3].

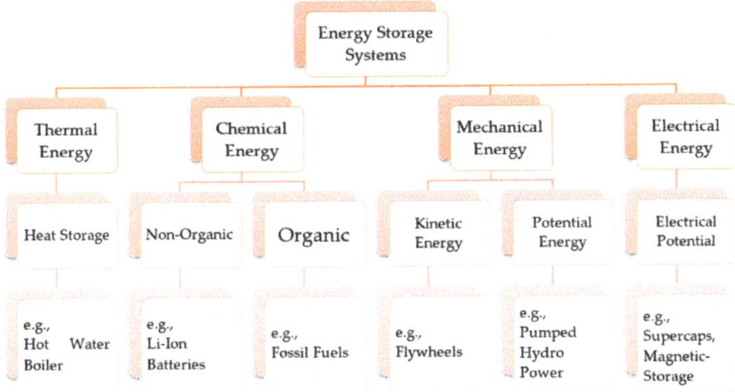

Figure 1. Classification of energy storage systems according to energy type, including examples.

A further classification is made in Figure 2, where different energy storage types are shown as a function of their power rating, energy content, and the consequently-related typical charge and discharge time. The dotted circle in the figure represents the area of particular importance for the use-case sustainable transportation, limiting the number of different storages to batteries, supercaps, flywheels and superconducting magnetic energy storages (SMES). However, the selection can further be reduced as SMES do not yet have the maturity necessary for a real implementation [4,5].

Beyond the obvious use-case of onboard energy storage, stationary buffer storage inside the required electric vehicle fast-charging stations will also be discussed in Section 3.3. Calculations and considerations are based on actual zero-emission buses operating in Graz, Austria. The main influencing parameters and effects related to energy storage aging are analyzed in detail. Based on the discussed aging behavior, advantages/disadvantages, and techno-economic analysis for both use-cases is presented. A final suitability assessment of each energy storage technology concludes the use-case analysis.

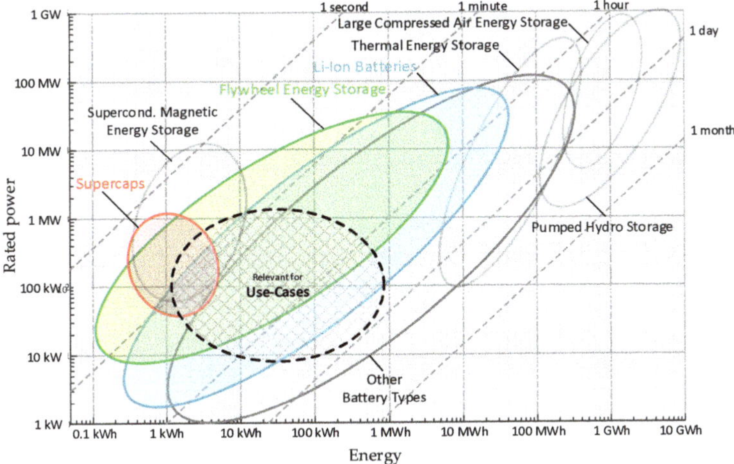

Figure 2. Power rating, energy capacity and discharge time of different energy storage systems for stationary and mobile transportation applications. Data based on References [6,7].

2. Properties of Different Energy Storage Systems

To give an overview, Table 1 shows general technical and economic properties of the storage technologies preselected in Section 1.

Table 1. General characteristic and economic properties of different energy storage technologies relevant to sustainable transport applications.

Energy Storage Technology	Specific Energy (Wh/kg)	Specific Power (kW/kg)	Temp. Range (°C)	Cycle life (-)
Mechanical Energy Storage				
Flywheel Energy Storage Systems (FESS)	10–100	>10	−30 °C to +70 °C	limitless
Chemical Energy Storage (only Li-Ion Batteries)				
Lithium Iron Phosphate (LFP)	90–120	4	−20 °C to +60 °C	5000–6000
Lithium Titanate LTO	60–80	1	−30 °C to +75 °C	>15,000
Lithium Nickel Cobalt Aluminum oxide (NCA)	200–300	1	−20 °C to +60 °C	500
Lithium Cobalt Oxide (LCO)	150–200	1	−20 °C to +60 °C	500–1000
Lithium Manganese Oxide (LMO)	100–150	4	−20 °C to +60 °C	300–700
Lithium Nickel Manganese Cobalt oxide (NMC)	150–280	1–4	−20 °C to +55 °C	3000–4000
Electrical Energy Storage				
Double Layer Capacitor (DLC)	5–10	>10	−20 °C to +60 °C	1 million
Hybrid Capacitor	10–20	>10	−20 °C to +60 °C	>50,000

Values refer to cell-level for batteries/supercaps and the rotor only for FESS, neglecting periphery and auxiliary systems. Some properties strongly depend on operating conditions, such as ambient temperature. Therefore, it must be mentioned that representative average characteristic values based on References [6,8–17] were used.

2.1. Battery and Superap

Both Li-Ion batteries and supercaps are mature technologies that have been used in various fields of application since the beginning of the 21st century. However, due to developments in recent years, Li-Ion batteries have become the energy storage device of choice for most transportation applications. Because of their popularity, a lot of scientific [6,7,11] and industrial [8–10,12–15], literature (provided by manufacturers) exists, which can be used to assess certain properties, such as cycle life, aging, etc. This paper will, hence, give only a short overview and primarily focus on the lesser-known properties of Flywheel Energy Storage Systems (FESS)—see Section 2.2.

When it comes to "usable/achievable lifetime", a metric is necessary to 'measure' the state of health of batteries. Typically, capacity and/or internal resistance are used in datasheets of cell manufacturers, giving some indicators to define the end-of-life (EOL) condition, e.g., a decrease of capacity by 20% or increase of internal resistance by a factor of two compared to the begin-of-life (BOL) values. In reality, those limits depend on the actual application, and the datasheet's lifetime values need to be scaled accordingly. Many applications allow a much higher decrease in capacity than defined by the manufacturer. This slightly increases initial costs and weight, but tremendously extends service life, e.g., a typically used value of 33% decrease of capacity results in a BOL to EOL capacity ratio of 1.5 compared to 1.25 for the manufacturer's 20% value. In this case, the battery would weigh (=cost) about 20% more. However, the lifetime would increase by about 65%. In other words, the battery would weight (=cost) less for a given lifetime and reach a higher over-all energy throughput. Only small benefits are gained by pushing it even further. Especially in transportation applications, the initial increase in weight is the limiting factor.

The achievable lifetime and performance of batteries and supercaps depend on many parameters, with temperature as the dominating influencing factor. Even though the values are given in Table 1 suggest a wide operating temperature range, a closer look into actual datasheets reveals the problems within: Temperature must be kept below a certain value in order to reach the highest cycle life. Table 2 illustrates the significant decrease in cycle life when temperatures exceed 25 °C.

Table 2. Temperature dependence of cycle life of an LFP cell from A123 (AMP20 [9]).

Effect of Temperature for 1 C/−C, 100% DOD Cycling for AMP20 Cells				
Cycle Count for Different Remaining Capacities at Different Temperatures				
Capacity	25 °C	35 °C	45 °C	55 °C
90%	2600	1450	850	400
80%	5150	3100	2000	850
70%	7700 *	5000	3050	1200

* Estimate, based on logarithmic extrapolation.

Low temperatures increase the internal resistance and thereby have a detrimental effect on the performance of the system as well. In the case of supercaps, even at the lowest allowed operating temperature, the increase is typically around factor two, e.g., the company AVX states an increase of about 120% at −40 °C compared to the reference value at 25 °C [8]. This decreased performance is still sufficient for common applications, and only the efficiency suffers slightly, but the capacity remains almost the same. Simultaneously, the (increased) losses heat up the supercap and thereby reduce the negative effects over time.

However, in the case of a battery, these effects are much more severe than in the case of supercaps. Typically, the temperature influence is already noticeable at around 10–20 °C. At temperatures below 0 °C, charging is often no longer allowed by the manufacturer. Finally, at the low end of the operating temperature range, the discharge performance of the cell is typically less than 10% compared to 20 °C values, e.g., References [9,13,15]. Due to this severe decrease in performance, it is often necessary to heat up the cells before the system is put into operation. According to the Graz Public Transport Services (GVB), putting a battery electric bus into operation in the winter may take up to 30 min.

The primary significance of high temperature is the decrease of the cell's lifetime, both calendar and cycle life. As a rule of thumb, one can assume that the calendar life is reduced by a factor of two every 10 °C increase in temperature (actual values taken from datasheets vary between 7 and 15 °C). The continuous operation at the maximum allowed temperature would reduce the lifetime to just a few months, or a year at most. An additional factor influencing the achievable lifetime is the cell voltage, and in the case of batteries also cycle count, depth of discharge (DOD), as well as charge and discharge rates.

For example, an AVX-SCC series supercap [8] has a base lifetime expectancy of 20 years at 30 °C in a fully charged state. The temperature coefficient is 8 °C per factor two in a lifetime, and the voltage dependence is 0.4 V per factor two in a lifetime (a lower voltage increases the lifetime, but reduces available capacity)—see Figure 3.

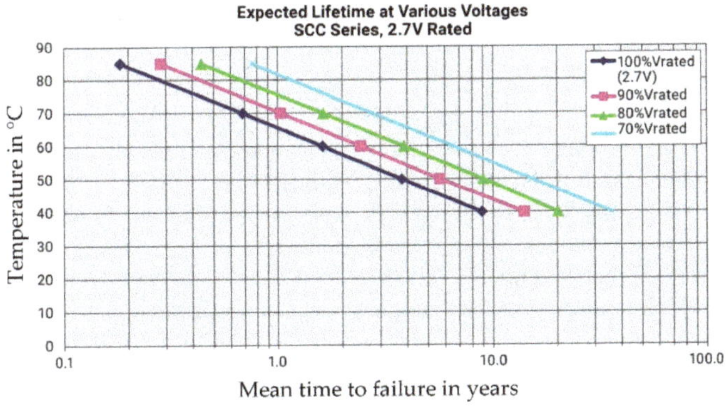

Figure 3. Expected supercap life depending on temperature and maximum voltage. (With kind permission by AVX Corporation) [8].

Similar to the supercap, calendar life of batteries depends on temperature and end-of-charge (EOC) voltage. Unfortunately, in many datasheets, only sparse data sets are given. If anything, one usually finds data regarding temperature dependence. Reasonable EOC voltages for different applications are rarely mentioned in datasheets/literature, with one notable exception being [15]. Depending on the application, Table 3 states different suggested EOC voltages. Still, it is not mentioned how much the lifetime is improved, due to voltage reduction.

Table 3. EOC-voltage depending on the application of a Li-Ion cell (Samsung—INR18650-35E) [15].

Standard Charging Voltage: Battery—4.20 V Cell (Samsung INR 18650-35E)			
Application	**EOC-Voltage**	**Application**	**EOC-Voltage**
Portable IT	4.20 V	E-Bike/E-Scooter	4.10 V
Power-Tool	4.20 V	Electric Vehicle	4.10 V
Medical	4.10 V	Energy Storage	4.00 V

Additionally, cycling the battery reduces its lifetime. There is no simple correlation between charge/discharge cycles and occurred damage. As mentioned before, it not only depends on the cycle count, but among other factors, also state-of-charge (SOC), DOD and charge/discharge rates. Still, a few basic and generally valid statements can be made:

- Just like calendar life, cycle life is influenced by cell temperature, but not necessarily with the same temperature coefficient. e.g., in Reference [9] the calendar life temperature coefficient is 10 °C per half/double lifetime, but for cycle life, the coefficient is 14 °C.
- Increasing charge/discharge rates reduce cycle life. High cycle life values, as shown in Table 1, are typically obtained by utilizing low charge/discharge rates, e.g., 1 C (1 h charge/discharge rate) or even lower. Increasing these rates, as often necessary for high-speed charging or other heavy-duty applications, reduces lifetime. Especially when the cell is optimized for high specific energy content, which is mainly the case for most batteries used in electric vehicles, where weight is of major interest.
- DOD influences the achievable energy throughput [10,14]. For example, Saft Evolion (NCA chemistry) reaches a cycle life of 4000 cycles for a DOD of 100% [14]. For a DOD of 10%, the cycle life increases to 250,000, resulting in a total energy throughput equivalent to 25,000 100%-cycles.

One last comment: In the case of rectangular or pouch bag cells, special care has to be taken for correct mounting that homogeneously compresses the cell with a defined pressure. This is equally important for the proper functioning of the cell, as well as to achieve long cell life.

2.2. Flywheel Energy Storage Systems (FESS)

2.2.1. Background Information

Prices of Lithium-Ion batteries are decreasing on the global market and energy densities have reached reasonable values, allowing EVs to travel 200 km and more on one charge [18]. However, there are still significant technical challenges, which need to be solved, or alternatives need to be found. One of the major drawbacks of chemical batteries is limited cycle life, which was described in Section 2.1 and will be discussed in particular in this paper.

It must be stressed that sustainable transportation does not only rely on batteries inside the vehicles. The increasing primary electricity supply through volatile sources, in combination with high grid loads caused by charging power demand, requires decentralized electric energy storage [19]. The requirements for these stationary energy storage systems may differ significantly from those of transportation applications. However, in both cases, long cycle life and negligible aging effects are usually desired. This is particularly the case when alternatives to chemical batteries come into play.

One may think immediately of gyroscopic reactions as a major disadvantage of FESS. This aspect must be considered during system design, but is an issue that can be resolved [20] as they have been used successfully in various transportation applications (see Figure 4).

(a) (b)

Figure 4. (a) The famed Gyrobus by Maschinenfabrik Oerlikon in 1953 powered solely by a 1500 kg electromechanical steel flywheel. (Image credit by Historisches Archiv ABB Schweiz, N.3.1.54627); (b) A modern transit bus accommodating a hybrid drive train with a flywheel energy recovery system by *PUNCH Flybrid*. (Image credit PUNCH Flybrid Ltd., Silverstone, UK).

Flywheel Energy Storage Systems (FESS) has experienced a renaissance in recent years, mainly due to some of their intriguing properties:

- In principle, an unlimited number of charge/discharge cycles;
- No capacity fade over time;
- Power and energy content are independent of each other;
- Operation at low or elevated temperature is easily possible;
- Precise state of charge (SOC)/state of health (SOH) determination;
- No risk during transportation/uncritical deep-discharge (flywheel stands still);
- No toxicologically critical/limited resources necessarily required.

Due to the above-listed properties, FESS are increasingly used for grid stability or fast-charging applications, as proposed in References [21,22]. Another example is the currently ongoing Austrian research project "FlyGrid", within which a FESS for a fully automated EV charging station will be developed. One module of this prototype will be used as the reference case and will deliver 5 kWh at 100 kW peak power.

2.2.2. FESS Working Principle

In a FESS, energy is stored in kinetic form; the working principle is based on the law of conservation of angular momentum. In electromechanical FESSs an external torque is applied to a rotor by the use of a motor/generator, hence, only an electrical and no direct mechanical connection for power transmission is required. In order to charge the FESS, the applied torque accelerates the spinning mass (rotor). If the spinning mass decelerates, energy is taken out of the system, and the motor acts as a generator. Electrical energy from the grid or other sources can be converted into kinetic energy charging the FESS. In the case of discharge, the motor/generator decelerates the spinning mass converting kinetic energy back to electrical energy. This principle is demonstrated in a video in the

Supplementary Materials, that belongs to this publication. The amount of stored energy is defined by the rotor's moment of inertia and the rotational speed, according to Equation (1).

$$E_{KIN} = \frac{I * \omega^2}{2} ,$$
(1)

E_{KIN} Kinetic Energy in J

I Mass Moment of Inertia of the Spinning Mass/Rotor in kg*m^2

ω Angular Velocity in rad/s

Different concepts for Flywheel Energy Storage Systems (FESS) exist, but within this publication, only electromechanical FESS are considered, as they are easily comparable to any other energy storage system with electric connection terminals. Figure 5a shows a schematic diagram of an electromechanical FESS. It consists of a motor/generator with a shaft and an attached spinning mass. The shaft is supported by bearings, which form the connection to the housing. In Figure 5b, the energy content is plotted over rotational speed visualizing the quadratic increase of stored energy. If high specific energies are desired, FESS must operate at extremely high rotational speeds, optimally exploiting rotor material strength. Within this publication, only FESS with high specific energies is addressed. As mentioned in Section 2.2.1 depth of discharge (DoD) does not influence FESS cycle life.

One effect, which must be considered during FESS design is based on Equation (2)—System power is proportional to motor torque and rotational speed. It can be observed that, when the constant power output is required at low rotational speeds, motor-generator torque will reach unnecessarily high values, resulting in heavier and more expensive electric machines. This is why the minimum operating speed is usually kept at around 1/3 of the maximum rpm value.

$$P = M * \omega ,$$
(2)

P Power in W

M Motor Torque in N*m

ω Angular Velocity in rad/s

Around 89% of the total kinetic energy of the FESS is usable when the system is operated between 33 and 100% of the maximum permissible speed. For that reason, FESS usually operate within a certain bandwidth and do not decelerate down to standstill during regular operation.

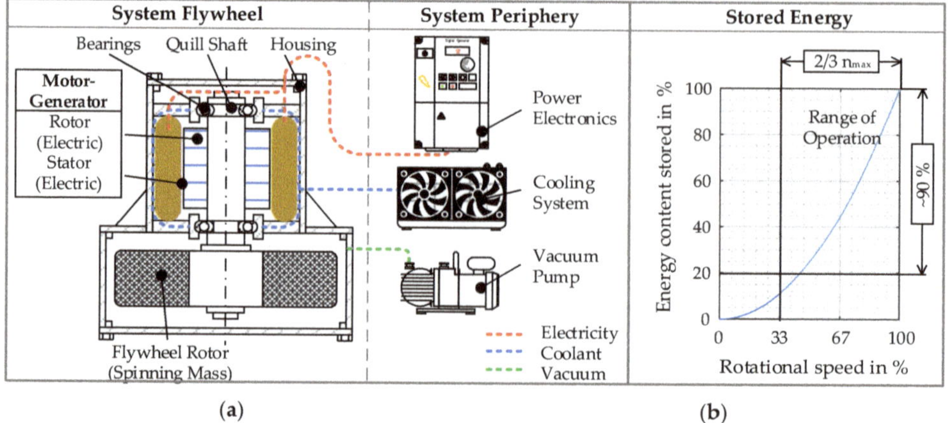

Figure 5. (**a**) Schematics of a flywheel energy storage system, including auxiliary components; (**b**) Energy content as a function of rotational speed.

2.2.3. Self-Discharge of FESS

Losses have an important influence on the suitability of this technology for different use-cases. The following paragraph gives a short introduction to this topic, starting with the three main causes of FESS self-discharge:

- Air drag;
- Bearing torque loss;
- Power consumption of peripheral components.

Air drag losses during operation are crucial for FESS with high specific energies. Circumferential speeds beyond the speed of sound are common and exceed 1 km/s in some cases, which would cause enormous air drag during operation. This air drag would result in losses and eventually be dissipated into heat, which causes thermal issues leading to system failure. In order to reduce these losses, FESS are usually operated in a vacuum atmosphere, and pressure levels down to 1 µbar are common [23]. At such low-pressure levels, air drag losses play only a minor role. Issues regarding lubrication arising from these vacuum qualities will be addressed in Section 2.2.6. The power consumption of the vacuum pump and other peripheral components must be taken into account when analyzing the overall system losses. The power dissipated in the bearings also plays a crucial role and will be discussed in detail in Section 2.2.5.

Losses do not necessarily represent a problem, when they are below a certain level, but the benchmark for this threshold depends on the actual use-case. With increasing mean power-transfer into and out of the FESS, the acceptable level of system losses increases as well.

This means that for long term storage (low mean power-transfer), the power loss threshold is very low and FESS is not suitable, due to its relatively high self-discharge (hours to days at most). For highly dynamic and predictable load cycles with high mean power-transfer FESS is more suitable. This matter will be demonstrated in Section 3 by means of different use-cases.

In the following sections, crucial FESS components will be dealt with, and details regarding their service life will be discussed.

2.2.4. Rotor Material Selection and Aging

As described in Section 2.2.1 FESS with high energy densities are addressed. Regarding the rotor, rotational speed, and therefore, energy content is limited by permissible stresses (σ_{max}) in the rotor. Highest energy densities can be reached when using materials with high $\frac{\sigma_{max}}{\rho}$ ratios, like fiber composite materials [1]. However, other materials like steel are being used in practice as well. Table 4 compares the theoretical specific energies of different rotor materials.

Table 4. Possible FESS rotor materials and associated theoretical specific kinetic energy content.

Material	Tensile Strength σ_{max}	Density ρ	Energy Density σ_{max}/ρ
	N/mm² (MPa)	Kg/dm³	Wh/kg
Mild Steel	340	7.8	12.1
Standard Electrical Sheet	400	8	13.9
Alloy Special Steel (*42CrMo4*)	1100	7.8	36.6
Birchwood	137	0.65	58.5
Aluminum ("*Ergal 65*")	600	2.72	61.3
Titanium ("*ZK 60*")	1150	5.1	62.6
High Strength Steel (*AlSi 4340*)	1790	7.83	63.5
Metal Matrix Composite	1450	3.3	122
Fiber reinforced plastic (E-Glass/EP 60%) *	960	2.2	132
Kevlar ("*Aramid 49EP*"/60%) *	1120	1.33	234
Carbon Fiber ("*M60J*") *	2010	1.5	372
Carbon Fiber ("*T1000G*") *	3040	1.5	563

* For composite materials a ratio of 60% fibers and 40% (matrix/resin) was assumed.

A spinning mass with a kinetic energy content of 5 kWh would weigh around 9 kg when made of carbon fiber reinforced plastic (CFRP) and 137 kg when 42CrMo4 high-strength steel is used, and the material is fully exploited regarding permissible stress. Commercially available systems reach specific energies regarding the rotor up to 50 Wh/kg when using CFRP, for steel flywheels, according to values are much lower.

However, it must be mentioned that the theoretical specific energy values are reduced by design parameters, such as safety factors, stress concentration (notching), etc. Figure 6 depicts the specific energy content of various real flywheel rotors with Li-batteries and fossil fuels. To show the enormous future potential of FESS technology, the theoretical specific energy potential of a rotor made from a material with properties similar to carbo nano-tubes is shown as well.

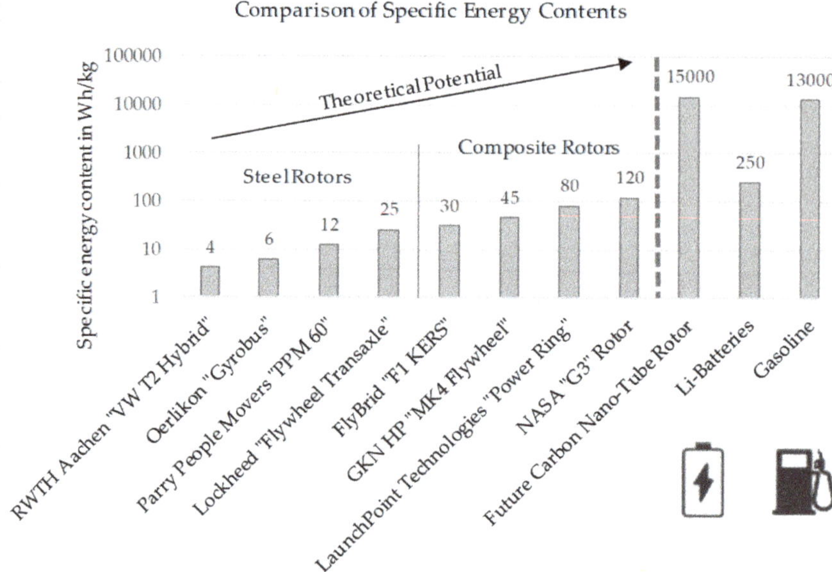

Figure 6. The specific energy content of selected real world flywheel rotors compared to future rotor potential and other energy storage [24].

Usually, the rotor weight is being compensated using a magnet (permanent magnetic thrust bearing), so that the ball bearings are not subjected to the entire rotor weight. The influence of rotor weight compensation is discussed in detail in Section 2.2.5.

While in theory FESS rotors are also subject to aging, due to fatigue stress (or even creep in the case of CFRP rotors) it must be mentioned that these phenomena are usually considered during the design phase by the introduction of a safety factor, and hence, do not result in the capacity fade of the system. In this regard, even rotors made of CFRP can reach high service life when they are designed accordingly, and aging of the matrix is considered [25].

2.2.5. Bearings

In most FESS, two fundamentally different bearing concepts are used: Active magnetic bearings (AMBs) and rolling element bearings (REBs). High costs compared to competing energy storage devices represent one of the major market entry barriers for FESS. For that reason, the upcoming sections focuses on low-cost solutions using REBs. Table 5 gives a brief overview of REBs compared to AMBs.

Table 5. Comparison of active magnetic bearing and rolling element bearing concepts.

Type	Active Magnetic Bearing	Rolling Element Bearing
Image		
Cost	Very High	Low
Stiffness	Low	High
Friction Losses	Very Low	High (Load and Speed-dependent)
Additional auxiliary energy supply	High	Not Required
Space requirement	High	Low
Lifespan	Very high	High (Load Dependent)

The REB's service life mainly depends on the applied loads. Generally, bearing loads are caused by rotor weight and machine dynamics/imbalance forces. There are different approaches to minimize bearing loads:

- Resilient bearing seat/supercritical rotor operation [26,27];
- Passive magnetic weight compensation [20,28];
- Precise rotor balancing [20,29];
- Active vibration control concepts [30].

Concepts based on active vibration control will not be considered within this publication as they have not reached readiness for marketing in FESS yet [26]. The other three measures are considered and explained briefly in the following paragraph using an example with the specific properties stated in Table 6:

Table 6. Properties of a reference FESS for parameter study regarding bearing life.

Specification of Reference System for Parameter Study *	
Energy Content	5 kWh
Peak Power	100 kW
Rotor Weight	130 kg
Bearing Configuration	Hybrid Spindle Bearings Myonic 30550 VA—Contact Angle: 15°, (X-Arrangement)
Mean Rotational Speed	20,000 rpm
Radial Bearing Load due to Imbalance (Reference)	100 N
Magnetic Rotor Weight Compensation (Reference)	95% (= 65 N axial load)

* Reference FESS module specifications based on the research project FlyGrid.

Axial bearing loads can, in fact, be almost entirely compensated by a passive magnetic lifting system, as shown in Figure 7, provided the system is designed with a vertical axis of rotation, which is normally the case. As shown in Figure 8a weight compensation is key to reach reasonable bearing service life. An attracting configuration using a ring magnet that would directly pull a ferromagnetic steel element on rotor upward has some disadvantages because of high eddy current losses at high rotor speeds. Using an additional permanent magnet on the rotor acting as a counter pole, the eddy current losses can be almost eliminated. For the remaining decision of either pulling the rotor on top or pushing it upwards from magnets mounted at the bottom, the latter configuration is preferable, due to its inherent stability of this configuration taking into account the direction of gravity. A demonstration video showing this configuration is uploaded in the Supplementary Materials of this publication.

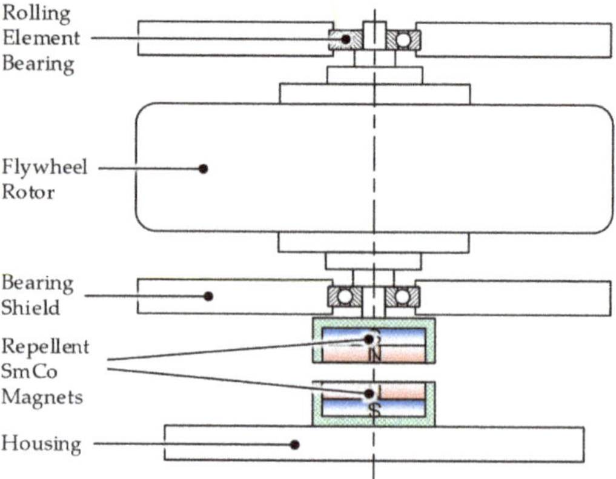

Figure 7. Cartoon image of low-cost, low-loss bearing configuration, including passive magnetic rotor weight compensation.

However, even if the rotor weight is nearly entirely compensated, imbalance forces remain. Based on a constant imbalance force of 100 N, bearing life for different compensation levels is shown. Without weight compensation, the bearings last only 21 days, but when 95% of the weight is compensated, bearing life increases up to more than 90 years. It must be noted that at compensation levels above 100% one bearing might be completely relieved which may cause slippage of the balls and lead to rapid system failure [31].

A flexible bearing suspension is used to operate the rotor supercritically. This means that at least the first two eigenfrequencies are surpassed and "self-centering" of the rotor occurs. During supercritical rotor operation bearing primarily loads, depend on the bearing seat's stiffness and the rotor imbalance and not on rotational speed [26]. Figure 8b shows the decrease of bearing life from 90 years to 25 years, when the rotor imbalance force is increased from 100 N to 300 N. Higher imbalance requires higher axial prestress of the bearing configuration, which is taken into account. For this study a magnetic weight compensation of 95% resulting in a remaining weight load of 74 N is assumed. Still, it must be mentioned that rotor imbalance may change over time, due to creep, wear or setting of joints.

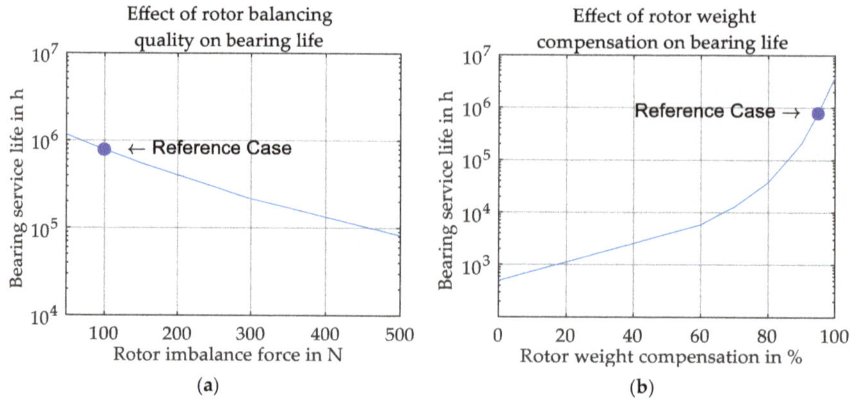

Figure 8. Influence of weight compensation (**a**) and rotor balancing quality (**b**) on bearing life (based on the reference case, shown in Table 6.).

Bearing friction torque is one of the main causes of losses in FESS and is mainly influenced by the following parameters:

- Size and geometry of the bearing
- Applied loads
- Rotational speed
- Lubrication

In order to calculate the resulting power loss bearing friction torque must be multiplied with the rotational speed.

The influence of bearing size, cage material and weight compensation factor on torque loss is a very complex matter and outside the scope of this publication. Detailed studies on the minimization of FESS bearing losses are available in References [1,32,33]. Still, as lubrication strongly affects FESS service life, the effects of lubricant viscosity and minimum quantity lubrication are discussed in the subsequent Section 2.2.7.

2.2.6. Lubrication

Generally speaking, there are three different lubrication principles for high speed rolling element bearings in FESS:

- Grease
- Oil
- Solid lubrication

For the considered use-cases are only oil and grease lubrication are relevant, as solid lubrication is mainly used when ambient pressures below 0.1 to 0.01 µbar are required [34]. Furthermore, oil and grease have superior service life for application in FESS compared to solid lubrication, and both are applicable for the considered use-cases. It must be noted that special vacuum grease/oil must be used in order to avoid outgassing, which has a detrimental effect on vacuum quality and may even lead to system failure.

Grease lubrication represents the most commonly used lubricating concept because it requires the least constructive and financial effort to implement, but shows some drawbacks regarding service life compared to oil lubrication [35]. Usually, fresh grease is stored in cartridges and extracted on demand. Standard recommended shelf life for grease in closed and sealed cartridges goes up to five years when stored properly [36]. During operation, service life depends strongly on applied loads and temperature. Above the permissible continuous maximum temperature for a specific grease, a temperature rise of 15 °C cuts grease service life in half [32]. Therefore, thermal management is crucial for FESS. In order to reach high service life, ongoing maintenance and grease change is required.

Oil lubrication is superior to grease with respect to service life. The used oil can be filtered continuously, thermally conditioned and may easily be changed on demand. Initial costs and effort to implement an oil lubrication circuit are significantly higher compared to grease. Due to the operation under vacuum, the lubrication concept must meet special requirements. Though the functionality of oil lubrication in FESS has been demonstrated in various research projects [37], these systems are not available off-the-shelf.

2.2.7. FESS Service Life

Based on the reference system presented in Table 6, Figure 9 summarizes the main influencing factors and their effect on bearing and lubrication service life.

In this example, grease lubrication and operation at the maximum continuous temperature limit are assumed. Therefore, every increase of 15 °C decreases lubrication service life by 50% until the maximum permissible operating temperature is reached (not shown in Figure 9). In short, FESS service life > 25 years is feasible with only minor maintenance effort.

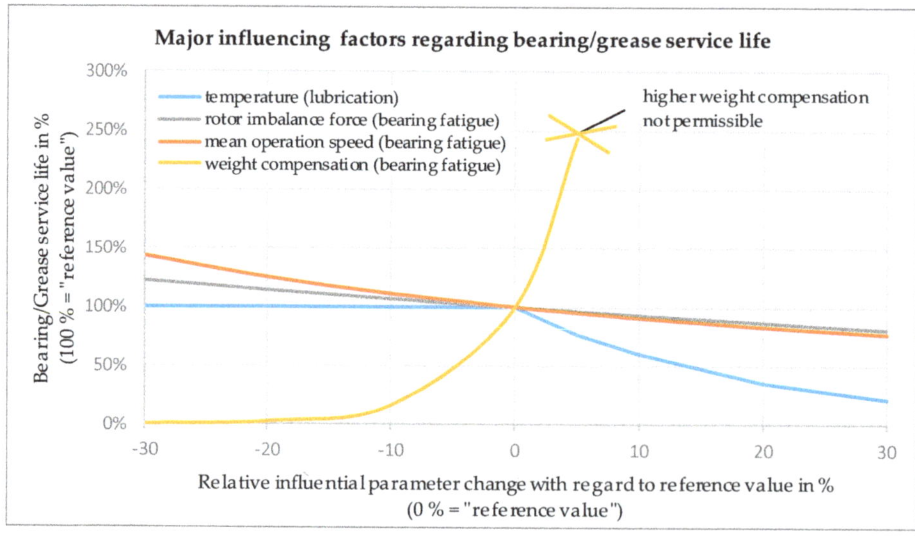

Figure 9. Major influencing factors (temperature, imbalance force, operational speed, weight compensation) on bearing and lubrication service life for the reference case listed in Table 6.

3. Use-Case Analysis and Results

In order to dimension any energy storage device accordingly, the predictability of the duty cycle is absolutely essential. Hence, one of the best predictable use-cases—public transport bus service—was chosen. To be more precise, the case of fully electrified buses with on-board energy storage (no hybrid- or trolley-variants) is considered. To specify the requirements for the energy storage onboard the bus, or possibly inside the charging station as buffer storage for grid load mitigation, the city of Graz in Austria is used as an example:

According to the public transport company (https://www.holding-graz.at/graz-linien) of Graz in Austria, 151 buses (currently mainly equipped with diesel engines) drive an average of 25,000 km per day. In total each bus travels around 0.5–1 million kilometers during its expected lifetime of 10–15 years. Using a representative urban city bus route ("Route 63" in Graz), the typical energetic requirements for a 12-m-bus were derived based on experiences gathered by the operator, and are shown in Table 7.

Table 7. Energetic properties of the reference use case for energy storage analyses.

Average Values Used as Reference Based on a Representative Bus Route ("Route 63" in Graz)			
Daily Mileage per Bus	150–200 km	Typical Round Trip Duration	1 h
Expected Lifetime	10–15 years	Average Energy Consumption	1.5 kWh/km *
Typical Round Trip Distance	12 km	Energy Consumption per Loop	18 kWh

* ... Value based on operator experience, including energy demand for heating and cooling.

Operating time during a whole day is around 16–18 h, so around 300 kWh storage would be needed to drive a whole day without any charging stops. For the most part of the day, there are six buses on the route simultaneously. Therefore, a bus arrives at the 'end-stop' every 10 min and (depending on traffic) a few minutes are left before it has to leave again. This time (0–5 min) could be used as a 'charging-window'. Assuming a 2-min charging window at the 10 min end stop (The actual duration of the end stop may vary depending on traffic; hence, a minimum of 2 min is assumed for charging.), an average power of 540 kW would be needed to transfer the previously calculated 18 kWh. Figure 10 shows two electric buses during operation in Graz.

(a)   (b)

Figure 10. Fully electric buses operating in Graz, Austria: (**a**) Bus equipped with supercaps at the pantograph charging station (Route 50); (**b**) Bus with Lithium-Ion batteries (Route 34).

On-board Energy Storage:

In principle, three different scenarios for the in-bus (onboard) energy storage have to be considered:

1. 'large energy storage'—the energy storage capacity is large enough, so that the bus could travel the whole day without recharging. The bus may be charged in the bus garage overnight, or partially at the end-stop during the day.
2. 'medium energy storage'—the bus can travel a few round trips without recharge, but not the whole day. The bus can be charged at the end-stop only, or partially at the end-stop—slowly depleting the storage during the day—followed by a full charge overnight in the bus garage. The rush-hour can be handled without recharging the storage, and thus, the timetable can be met easily. Compared to scenario 1, this concept offers some advantages regarding weight and storage costs.
3. 'small/minimum energy storage'—the capacity is enough for a single loop, and the bus has to be recharged every time at the end-stop. However, the energy storage size depends on the route (travel distance, duration), weather conditions (heating/cooling energy need), traffic conditions (especially congestion), etc. making it nearly impossible to design a universally applicable cost-effective solution. Today, typically a combustion engine is added to the vehicle for this scenario leading to a hybrid electric vehicle (HEV) solution.

Due to the limitations of the 'small/minimum storage' scenario and the exclusion of hybrid solutions, for further calculations, only case 1 and 2 are taken into account.

Charging Station Energy Storage:

In terms of the integration of renewables into the grid, two aspects are important:

- The bus-recharge should be performed during the day, due to solar power availability. (Even though a bus route in Austria was selected and Austria has a lot of hydro power, the general use-case is relevant to any country, and since the global solar power potential is higher than the present world energy consumption [38], the solar peak should be considered)
- A uniform grid load should be pursued to avoid thermal overload, instabilities, voltage drop, etc.

This can be achieved by utilizing an energy storage device located directly at the charging station, as it was proposed and tested in References [11,16,39] and depicted in Figure 11.

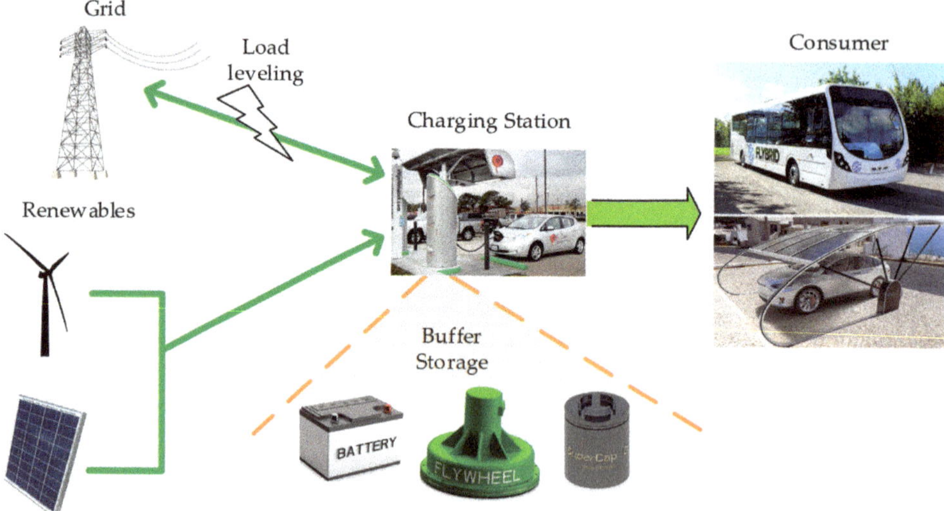

Figure 11. Electric vehicle charging station concept with built in buffer storage. (Photos with kind permission by SECAR Technologie GmbH and PUNCH Flybrid).

Given the previous example of 2-min-long 540 kW charge pulses every 10 min, an energy storage enhanced charging station with 90% efficiency would smooth out the power drawn from the grid to a continuous value of around 130 kW. This would lower the demand on the grid by avoiding possible (thermal) overloads of the powerlines and keeping voltage and frequency within the required specifications. During an entire day, around 100 charging processes are performed accumulating to about 35,000 cycles per year. An additional advantage of storage enhanced charging station is the reduced grid power compared to overnight charge: A single bus needs about 300 kWh per day. Therefore, charging all six buses in about 5 h overnight needs about 360 kW—around three times more compared to the previously calculated 130 kW.

The high demands regarding power, cycle life and energy content of energy storage for professional use in public transportation require careful evaluation of the underlying aging processes, as discussed in Section 2. For the considered on-board use-cases ('medium' and 'large' energy storage) supercaps and flywheels were ruled out for the following reasons:

(a)   The high required energy makes supercaps too heavy and expensive.
(b)   FESS have major drawbacks regarding self-discharge behavior, as well as cost and weight, which represent the main requirements for transportation applications.

3.1. On-Board 'Large' Energy Storage

For the given the example, a 300 kWh energy storage device is needed to allow operation for an entire day. To keep the weight within limits energy storage with high specific energy should be used, e.g., an NCA or NMC Li-Ion battery. Since the battery deteriorates during usage, and 300 kWh is needed at EOL, the BOL capacity must be accordingly higher. Using a capacity deterioration of 33% to define the EOL state, the BOL capacity must be increased to 450 kWh.

With respect to lifetime analysis, cycle life will be dealt with at first. If the battery is charged overnight, it is cycled only once a day, resulting in a high DOD each day. Over the duration of 10 to 15 years, this leads to about 3000–5000 cycles, or an energy throughput of 1–1.6 GWh. Normally, high energy cells are not capable of sustaining that many cycles. Typically, they are able to deliver only 500–1000 cycles (100% DOD), which for the 450 kWh battery would give an energy throughput of about 0.23–0.45 GWh until EOL. Therefore, they would have to be exchanged one or more times over the lifetime of the bus. However, if the operating regime is changed to intermediate charging at the end station, the DOD is significantly lowered. In the given use-case the energy transferred for a single cycle is 18 kWh. Compared to the battery capacity of 450 kWh this corresponds to a DOD of only 4%. As previously shown, the possible energy throughput during a battery's lifetime depends tremendously on the DOD—for some batteries, the manufacturers claim an inverse proportional behavior on DOD [10,14]. That means, for a DOD of 4% an increase of cycle life by a factor of 25 would be achievable, resulting in a possible energy throughput of 6–11 GWh, far exceeding the needed 1–1.6 GWh previously shown. So, even if the actual battery does not scale as well, a (cycle) lifetime of 10 to 15 years should still be feasible.

Of course, another possibility is to use cells with higher cycle life to enable the possibility of overnight charge. Those could either be a different battery technology (e.g., LTO or LFP), or they could still be of NCA or NMC type (but optimized for longer life). In both cases, the disadvantage would be a higher battery weight, due to lower specific energy of the cells.

Taking into account the calendar life as well, a significant additional decrease in capacity would occur during those 10–15 years. Hence, most likely, the battery would still need to be replaced after about 5–10 years. Still, to reach such high lifetime in this heavy-duty application thermal conditioning to low operating temperatures (around 20 °C) is mandatory. Table 8 summarizes the above elaborations.

Table 8. Comparison of charging regimes for the use-case of 'large' on-board energy storage.

	Use-Case: On-Board 'Large' Storage	
	Battery (NCA or NMC—Cell): Intermediate Charging	Battery (NCA or NMC—Cell): Only Overnight Charging
Calendar Life [a]	20–25 years	20–25 years
Cycle life	12,500–25,000 4% DOD-cycles	500–1000 100% DOD-cycles
Service Life [b]	10–15 years	3.5–4.5 years
Required Capacity (BOL)	450 kWh	450 kWh
Charge	540 kW	75 kW

[a] how long the storage is expected to last in terms of calendar years. Excludes influence of charge/discharge. [b] how long the storage is expected to last in the real application, including influences of charge/discharge and device temperature based on an EOL capacity fade of 33%.

3.2. On-Board 'Medium' Energy Storage

Using energy storage with less capacity can save cost and weight. For the example considered, a BOL capacity of 90 kWh (80% reduction in respect to the previous example) is assumed. Given the recharge power of 540 kW, this corresponds in a charging C-rate of 6, too high for a 'high energy' optimized battery. Moreover, since the energy storage has less capacity than in the above example, the cell must have a much higher cycle life, as is the case with LFP and LTO cells. These cells have less specific energy, reducing the possible weight savings, but still, an improvement by a factor 2–3 would be possible compared to the 'large' energy storage case.

Using an AMP20m1HD-A cell as an example (LFP cell from A123 [9]), the cell loses 10% capacity after 2700 cycles (100% DOD, 23 °C, 5 C charge- and 1 C discharge-rate). Since the DOD of the considered use-case is only 20% (18 kWh/90 kWh), it can be assumed that after an energy throughput of 1–1.6 GWh (cycling during 10–15 years) the cell has lost about 10–15% capacity. Due to the calendar aging additionally about 15–25% capacity is lost after 10–15 years (at 23 °C), resulting in about 25–40% overall capacity loss. Still, even the 40% loss after 15 years (with ~54 kWh remaining) would enable the bus to drive 3 whole rounds. Table 9 summarized the key findings explained in the paragraph above.

Table 9. Summary for the use-case 'medium' on-board energy storage.

Use-Case: On-Board 'Medium' Storage	
Battery (LFP Cell)	
Calendar Life [a]	10–15 years
Cycle life	8000–10,000 cycles
Service Life [b]	10–15 years
Required Capacity (BOL)	90 kWh
Charge	540 kW

[a] how long the storage is expected to last in terms of calendar years. Excludes influence of charge/discharge.
[b] how long the storage is expected to last in the real application, including influences of charge/discharge and device temperature based on an EOL capacity fade of 33%.

3.3. End-Stop Charge Station Energy Storage

In contrast to the on-board energy storage, which is charged once a day or every hour at the bus-stop, the charge station's energy storage is cycled every 10 min, resulting in about 35,000 cycles per year. Assuming a charge transfer efficiency of 90%, during the charge duration of 8 min 127 kW are drawn from the power grid, charging about 15 kWh into the energy storage. Afterwards, the energy storage is discharged within 2 min with a power of 413 kW. With the additionally 127 kW still drawn from the grid, the bus is charged with the desired 540 kW, as shown in Figure 12.

Using a chemical battery as buffer stationary storage in this scenario, a technology with very high cyclability (e.g., LTO) is necessary. For example, a 100 kWh battery built with Altairnano's LTO technology [10,40] would have 25 years calendar life or sustain about half a million cycles at 15% DOD (both values at 25 °C). In combination, this would give about 10–15 years of service life. Using supercaps could increase this interval to about 15–25 years [8]. However, like for the on-board energy storage scenarios, the thermal conditioning to 20–25 °C (max.) is absolutely necessary to achieve those long lifetimes for electrochemical energy storage.

Alternatively, using a FESS could tremendously extend the achievable service life, well beyond 25 years. Of course, there are some components, which may need to be replaced during the systems life span, like grease and oil in the case of mechanical bearings, the necessary maintenance of the vacuum pump or the exchange of electronic components. However, the main parts (flywheel mass, housing, electric motor), which also represent the main cost factors of the energy storage system are characterized by excellent longevity and do not need to be replaced. An additional advantage of FESS compared to electrochemical energy storage systems is the insensitivity to temperature—especially regarding lifetime—enabling the usage of simple water cooling circuits instead of costly chilling systems.

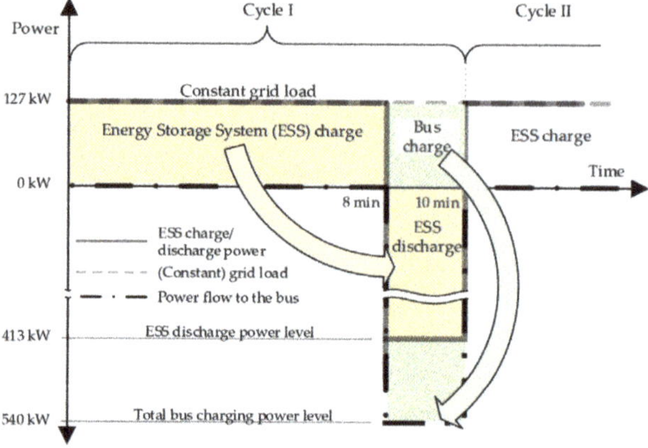

Figure 12. Grid load and charging power level of the load cycle for the "end-stop charge station use-case" shown for two consecutive charging cycles.

4. Summary

Within this study, energy storage for sustainable transport applications was investigated with respect to service life. The theoretical background of different energy storage systems, as well as different use-cases were described in detail. For exemplary energy storage comparison and benchmarking, the use-case of a fully electric transit bus operating in urban public transportation was selected, whereas onboard energy storage and buffer energy storage inside the fast-charging station were considered. From a great variety of options, three established and feasible energy storage systems were chosen for a more detailed analysis. For energy storage inside the fast-charging station, it was shown that high demand on cycle life and other requirements, such as short storage time, high power and long targeted service life clearly favor flywheel energy storage systems (FESS) over supercapcitors or batteries. However, fewer load cycles and long-time storage onboard the transit bus calls out for state of the art Li-Ion batteries rather than supercaps or FESS. Hence, which energy storage technology is most suitable strongly depends on the envisioned use-case and consequently the actual duty cycle.

A representative FESS module (5 kWh, 100 kW peak) was analyzed for the proposed use-cases and measures to improve/maximize service life was suggested. For the same use-case, a comparable battery system was analyzed, showing that battery size, and therefore, DOD (depth of discharge) is crucial for battery life. Beyond that, the battery solution requires accurate thermal conditioning and monitoring as calendar and cycle life are strongly affected when operated outside a narrow temperature band.

In this context, the great potential of FESS was shown. It is to be expected that in future some of the major advantages of FESS will be exploited, e.g., nearly unlimited cycle and calendar life, easy state of charge determination, independence from limited resources, etc. FESS-specific drawbacks, such as self-discharge, weight and cost may be detrimental to mobile (onboard) applications, but can be mitigated or neglected in some stationary applications, such as EV charging stations. Hence, FESS represent a valuable contribution to the energy revolution by increasing grid stability and facilitating the integration of renewables into the grid.

Supplementary Materials: The following are available online at http://www.mdpi.com/2071-1050/11/23/6731/s1, Video S1: passive magnetic weight compensation demonstrator, Video S2: Basic principle of flywheel energy storage.

Author Contributions: Conceptualization, P.H., A.B. and H.W.; Writing—Original Draft Preparation P.H.; Methodology, P.H., H.W. and B.S.; Software, Validation, Formal Analysis and Investigation, P.H. and B.S.; Data Curation, M.B.; Visualization, P.H., M.B. and A.B.; Project Administration, A.B.; Supervision, Funding Acquisition and Resources, H.W.

Funding: This research was conducted within the Project FlyGrid, funded by the Austrian Research Promotion Agency (FFG) within the Electric Mobility Flagship Projects, 9th call, grant number 865447. Open Access Funding by Graz University of Technology.

Acknowledgments: The authors would like to thank Myonic GmbH for their support regarding the service life of rolling element bearings and their lubrication, as well as Holding Graz for their input regarding urban bus fleet operation.

Conflicts of Interest: The authors declare no conflict of interest.

Abbreviations

AMB	Active Magnetic Bearing
BOL	Begin of Life
CFRP	Carbon Fiber Reinforced Plastic
DLC	Double Layer Capacitor
DOD	Depth of Discharge
EOC	End of Charge
EOL	End of Life
EV	Electric Vehicle
ESS	Energy Storage System
FESS	Flywheel Energy Storage System

GVB	Graz Public Transport Services
HEV	Hybrid Electric Vehicle
IT	Information Technology
Li-Ion	Lithium Ion
LCO	Lithium Cobalt Oxide
LFP	Lithium Iron Phosphate
LMO	Lithium Manganese Oxide
LTO	Lithium Titanate
NCA	Lithium Nickel Cobalt Aluminum Oxide
NMC	Lithium Nickel Manganese Cobalt Oxide
REB	Rolling Element Bearing
SMES	Superconducting Magnetic Energy Storage
SOC	State of Charge
SOH	State of Health

References

1. Buchroithner, A. *Schwungradspeicher in der Fahrzeugtechnik*; Springer: Berlin/Heidelberg, Germany, 2019.
2. MacKay, D. *Sustainable Energy—Without the Hot Air*; UIT Cambridge Ltd.: Cambridge, UK, 2009.
3. Patrick, J.G.; Moseley, T. *Electrochemical Energy Storage for Renewable Sources and Grid Balancing*; Elsevier Ltd.: Amsterdam, The Netherlands, 2014.
4. Xue, X.D.; Cheng, K.W.E.; Sutanto, D. A study of the status and future of superconducting magnetic energy storage in power systems. *Supercond. Sci. Technol.* **2006**, *19*, R31. [CrossRef]
5. Global Energy Network Institute (GENI). *Energy Storage Technologies & Their Role in Renewable Integration*; Springer International: Cham, Switzerland, 2012.
6. Luo, X.; Wang, J.; Dooner, M.; Clarke, J. Overview of current development in electrical energy storage technologies and the application potential in power system operation. *Appl. Energy* **2015**, *137*, 511–536. [CrossRef]
7. Weber, E.R. Frauenhofer-Institute for Solar Energy systems ISE and Albert Ludwigs University Freiburg, Energy Storage Technologies on the Way to the Global Market. In Proceedings of the 8th International Renewable Energy Storage Conference IRES 2013, Berlin, Germany, 18–20 November 2013.
8. AVX Corporation. High Capacitance Cylindrical Super Capacitors, SCC Series Super Capacitors. 2019. Available online: http://datasheets.avx.com/AVX-SCC.pdf (accessed on 4 October 2019).
9. A123 Energy Solutions. Battery Pack Design, Validation, and Assembly Guide Using A123 Systems Nanophosphate Cells, Document No. 493005-002. Available online: https://www.buya123products.com/uploads/vipcase/48ccae4db85064588e3d82c105ab4247.pdf (accessed on 4 October 2019).
10. Altairnano. PowerRack Description. 2019. Available online: https://altairnano.com/products/powerrack (accessed on 4 October 2019).
11. Makohin, D.; Viveiros, F.; Zeni, V.S. Use of lithium iron phosphate energy storage system for EV charging station demand side management. In Proceedings of the IEEE 8th International Symposium on Power Electronics for Distributed Generation Systems (PEDG), Florianopolis, Brazil, 17–20 April 2017.
12. Maxwell Technologies. Datasheet: 3.0V 3400F Ultracapacitor Cell—BCAP3400 P300 K04/05. Available online: https://www.maxwell.com/images/documents/3V_3400F_datasheet.pdf (accessed on 2 October 2019).
13. Panasonic. NCR20700B Data Sheet. Available online: http://www.batteryspace.com/prod-specs/10873%20specs%20.pdf (accessed on 3 October 2019).
14. Saft. Evolion Li-Ion Battery—Technical Manual; Document No. 21880-2-0115. Available online: https://www.manualslib.com/manual/1132262/Saft-Evolion.html (accessed on 30 September 2019).
15. Samsung SDI. Specification of Product for Lithium-Ion Rechargeable Cell, Model Name: INR18650-35E. Available online: https://www.bto.pl/pdf/08097/INR18650-35E.pdf (accessed on 1 October 2019).
16. Zhao, H.; Burke, A. An intelligent solar powered battery buffered EV charging station with solar electricity forecasting and EV charging load projection functions. In Proceedings of the IEEE International Electric Vehicle Conference (IEVC), Florence, Italy, 17–19 December 2014.

17. Maxwell Technologies. Product Guide—Maxwell BOOSTCAP Ultracapacitors; Document No. 1014627.1. Available online: https://www.maxwell.com/images/documents/PG_boostcap_product_guide.pdf (accessed on 2 October 2019).
18. U.S. Department of Energy. Office of Energy Efficiency & Renewable Energy. Available online: Fueleconomy. gov (accessed on 4 December 2018).
19. Buchroithner, A.; Wegleiter, H.; Schweighofer, B. Flywheel Energy Storage Systems Compared to Competing Technologies for Grid Load Mitigation in EV Fast-Charging Applications. In Proceedings of the 27th International Symposium on Industrial Electronics (ISIE), Cairns, Australia, 13–15 August 2018.
20. Buchroithner, A.; Brandstätter, A.; Recheis, M. Mobile Flywheel Energy Storage Systems: Determining Rolling Element Bearing Loads to Expand Possibilities. *IEEE Veh. Technol. Mag.* **2017**, *12*, 83–94. [CrossRef]
21. Karrari, S.; Noe, M.; Geisbuesch, J. High-speed Flywheel Energy Storage System (FESS) for Voltage and Frequency Support in Low Voltage Distribution Networks. In Proceedings of the IEEE 3rd International Conference on Intelligent Energy and Power Systems (IEPS), Kharkiv, Ukraine, 10–14 September 2018.
22. Sbordone, D.; Bertini, I.; Di Pietra, B.; Falvo, M.C.; Genovese, A.; Martirano, L. EV fast charging stations and energy storage technologies: A real implementation in the smart micro grid paradigm. *Electr. Power Syst. Res.* **2015**, *120*, 96–108. [CrossRef]
23. Genta, G. *Kinetic Energy Storage—Theory and Practice of Advanced Flywheel Systems*; Butterworth & Co. Ltd.: London, UK, 1985.
24. Buchroithner, A.; Haidl, P.; Birgel, C.; Zarl, T.; Wegleiter, H. Design and Experimental Evaluation of a Low-Cost Test Rig for Flywheel Energy Storage Burst Containment Investigation. *Appl. Sci.* **2018**, *8*, 2622. [CrossRef]
25. Bai, Z.; Wang, J.; Ning, K.; Hou, D. Contact Pressure Algorithm of Multi-Layer Interference Fit Considering Centrifugal Force and Temperature Gradient. *Appl. Sci.* **2018**, *8*, 726. [CrossRef]
26. Haidl, P.; Zisser, M.; Buchroithner, A.; Schweighofer, B.; Wegleiter, H.; Bader, M. Improved Test Rig Design for Vibration Control of a Rotor Bearing System. In Proceedings of the 23rd International Congress on Sound and Vibration, Athens, Greece, 10–14 July 2016.
27. DellaCorte, C. Novel Super-elastic Materials for Advanced Bearing Applications. In Proceedings of the CIMTEC International Ceramics Conference, Montecatini Terme, Italy, 8–13 June 2014.
28. Buchroithner, A.; Haan, A.; Preßmair, R.; Bader, M.; Schweighofer, B.; Wegleiter, H.; Edtmayer, H. Decentralized Low-Cost Flywheel Energy Storage for Photovoltaic Systems. In Proceedings of the International Conference on Sustainable Energy Engineering and Application (ICSEEA), Jakarta, Indonesia, 3–5 October 2016.
29. Buchroithner, A.; Andrasec, I.; Bader, M. Optimal system design and ideal application of flywheel energy storage systems for vehicles. In Proceedings of the 2012 IEEE International Energy Conference and Exhibition (Energycon), Florence, Italy, 9–12 September 2012.
30. Zisser, M.; Haidl, P.; Schweighofer, B.; Wegleiter, H.; Bader, M. Test Rig for Active Vibration Control with Piezo Actuators. In Proceedings of the 22nd International Congress on Sound and Vibration, Florence, Italy, 12–16 July 2015.
31. SKF Gruppe. SKF Hochgenauigkeitslager der Reihe "Super-Precision Bearings". Available online: https://www.skf.com/binary/78-129877/0901d19680363d4d-13383_1-DE---Super-precision-bearings.pdf (accessed on 30 September 2019).
32. Schäffler Technologies AG & Co. KG. *Wälzlagerpraxis: Handbuch zur Gestaltung und Berechnung von Wälzlagerungen*; Vereinigte Fachverlage GmbH: Mainz, Germany, 2015.
33. Buchroithner, A.; Voglhuber, C. Enabling Flywheel Energy Storage for Renewable Energies—Testing of a low-cost, low-friction bearing configuration. In Proceedings of the 12th VDI-Fachtagung Gleit-und Wälzlagerungen, Schweinfurt, Germany, 27–28 June 2017.
34. Birkhofer, H.; Kümmerle, T. *Feststoffgeschmierte Wälzlager—Einsatz, Grundlagen und Auslegung*; Springer: Berlin/Heidelberg, Germany, 2012.
35. Schaeffler Technologies Ag & Co. KG. Schmierung von Wälzlagern. Available online: https://www.schaeffler.com/remotemedien/media/_shared_media/08_media_library/01_publications/schaeffler_2/tpi/downloads_8/tpi_176_de_de.pdf (accessed on 15 September 2019).

36. SKF. SKF Maintenance and Lubrication Products. Available online: https://www.skf.com/binary/21-163650/03000EN.pdf (accessed on 10 September 2019).
37. Rosseta Technik GmbH. Energiespeicher für das Straßenbahnnetz. Available online: http://www.rosseta.de/texte/bahnsr.pdf (accessed on 20 September 2019).
38. Trieb, F.; Schillings, C.; O'Sullivan, M.; Pregger, T.; Hoyer-Klick, C. Global Potential of Concentrating Solar Power. In Proceedings of the 15th SolarPaces Conference, Berlin, Germany, 15–18 September 2009.
39. Renewable Energy World. Energy Storage for EV Charging in Canada to Combat Range Anxiety. Available online: http://www.renewableenergyworld.com/articles/2017/07/energy-storage-for-ev-charging-in-canada-to-combat-range-anxiety.html (accessed on 26 July 2016).
40. Altairnano. 24 V, 60 Ah Battery Module Datasheet; Document No. 3336145 R2. 2012. Available online: http://altairnano.com/wp-content/uploads/2015/11/24V-60Ah-BATTERY-MODULE-Data-Sheet.pdf (accessed on 7 October 2019).

MDPI

St. Alban-Anlage 66

4052 Basel

Switzerland

Tel. +41 61 683 77 34

Fax +41 61 302 89 18

www.mdpi.com

Sustainability Editorial Office

E-mail: sustainability@mdpi.com

www.mdpi.com/journal/sustainability